Qijun Gu   Wanyu Zang   Meng Yu (Eds.)

Security in Emerging Wireless Communication and Networking Systems

First International ICST Workshop, SEWCN 2009
Athens, Greece, September 14, 2009
Revised Selected Papers

 Springer

Volume Editors

Qijun Gu
Texas State University-San Marcos
Department of Computer Science
601 University Drive
San Marcos, TX 78666, USA
E-mail: qijun@txstate.edu

Wanyu Zang
Meng Yu
Western Illinois University
Department of Computer Science
1 University Circle
Macomb, IL 61455, USA
E-mail: {w-zang; m-yu2}@wiu.edu

Library of Congress Control Number: 2009942991

CR Subject Classification (1998): C.2, D.4.6, K.6.5, E.3, G.2, I.1

ISSN   1867-8211
ISBN-10   3-642-11525-X Springer Berlin Heidelberg New York
ISBN-13   978-3-642-11525-7 Springer Berlin Heidelberg New York

springer.com

© ICST Institute for Computer Sciences, Social-Informatics and Telecommunications Engineering 2010
Printed in Germany

Typesetting: Camera-ready by author, data conversion by Scientific Publishing Services, Chennai, India
Printed on acid-free paper   SPIN: 12830046   06/3180   5 4 3 2 1 0

Preface

The First ICST International Workshop on Security in Emerging Wireless Communication and Networking Systems (SEWCN 2009) was held in Athens, Greece, September 14, in conjunction with SecureComm 2009. SEWCN 2009 was sponsored by the Institute for Computer Sciences, Social-Informatics and Telecommunications Engineering (ICST). The Workshop Chairs were Qijun Gu from Texas State University-San Marcos, USA, and Wanyu Zang from Western Illinois University, USA.

The workshop invited 20 researchers from academia and industry around the world in the areas of networking and security to form the Program Committee. The workshop received nine submissions and each submission received two or three double-blind reviews. The review process started on July 6 and ended on July 27. In all, 21 reviews were received. Based on the review scores and comments, seven papers with average score 0 or better were accepted for presentation and inclusion in the workshop proceedings.

The workshop emphasized new ideas for secure architectures and protocols to enhance the emerging wireless systems. The accepted papers cover topics on applied cryptography, key management, vulnerability analysis, privacy, authentication, and intrusion detection for emerging wireless systems. The papers were presented in two sessions, chaired by Nikolaos Preve from the National Technical University of Athens, Greece, and Theofilos Chrysikos from the University of Patras, Greece.

For the success of SEWCN 2009, we would like to thank the Program Committee and the authors for their work and contributions. We would like to thank the technical sponsors CREATE-NET and ICST for their support and organization. We would also like to thank all the people involved in the organization of the workshop, in particular, the main Conference Chair Peng Liu, the conference coordinators Gergely Nagy, Eszter Hajdu and Effie Makri, and the general Workshop Chair Reza Curtmola.

Organization

Workshop Chairs

Qijun Gu   Texas State University-San Marcos, USA
Wanyu Zang   Western Illinois University, USA

Technical Program Committee

Abderrahim Benslimane   University of Avignon, France
Raheem Beyah   Georgia State University, USA
Ioannis Broustis   University of California, Riverside, USA
Timothy Brown   University of Colorado at Boulder, USA
Alvaro Cárdenas   University of California, Berkeley, USA
Hakima Chaouchi   Institut Telecom, France
Charles Clancy   Laboratory for Telecommunications Sciences, USA
Sghaier Guizani   Qatar University, Qatar
Wolfgang Hommel   Leibniz Supercomputing Centre, Germany
Jiankun Hu   RMIT University, Australia
Salil Kanhere   University of New South Wales, Australia
Frank Kargl   Ulm University, Germany
Costas Lambrinoudakis   University of the Aegean, Greece
Timothy Newman   Virginia Tech, USA
Raphael Phan   Loughborough University, UK
Elmar Schoch   Ulm University, Germany
Gritzalis Stefanos   University of the Aegean, Greece
Jay-Evan Tevis   LeTourneau University, USA
Yang Xiang   Central Queensland University, Australia
Meng Yu   Western Illinois University, USA

Table of Contents

Table of Contents

Session 1

Session Chair: Nikolaos Preve

A Closed-Form Expression for Outage Secrecy Capacity in Wireless Information-Theoretic Security

Theofilos Chrysikos, Tasos Dagiuklas, and Stavros Kotsopoulos

Wireless Telecommunications Laboratory
Department of Electrical & Computer Engineering
University of Patras – 26500 Greece
{txrysiko,ntan,kotsop}@ece.upatras.gr

Abstract. This paper provides a closed-form expression for Outage Secrecy Capacity in Wireless Information-Theoretic Security. This is accomplished on the basis of an approximation of the exponential function via a first-order Taylor series. The error of this method is calculated for two different channel cases, and the resulting precision confirms the correctness of this approach. Thus, the Outage Secrecy Capacity can be calculated for a given Outage Probability and for a given propagation environment (path loss exponent, average main channel SNR), allowing us to estimate with increased precision the boundaries of secure communications.

Keywords: Wireless Information-Theoretic Security; quasi-static Rayleigh fading; Outage Secrecy Capacity; Taylor approximation; path loss exponent.

1 Introduction

Security remains an issue of utmost importance in wireless communications. For all the advances and breakthrough progress in both industry and academia, security still provides a fertile ground for extensive research and innovative solutions. It is imperative to begin with a brief overview of background work in the field of wireless security from an information-theoretic standpoint.

1.1 Background Work

Based on Shannon's definition of *perfect secrecy* [1], innovative research was carried out in the latter half of the 1970s, investigating the impact of the wireless channel on the boundaries of secure communications [2]-[4]. Both the main and the wiretap channel were considered to be Gaussian. This proved to be the first major setback in the ongoing research, due to the limitation that the average SNR of the main (legitimate) channel had to be greater than the average SNR of the wiretap channel (eavesdropper's channel) so that secure communication over the wireless interface would be guaranteed. To make matters worse, the lack of channel coding schemes at the time prevented researchers from coming up with a flexible and reliable solution to the situation at hand. Information-theoretic solutions for wireless security were seemingly brought to a quick ending, and the interest of public research was drawn towards

Q. Gu, W. Zang, and M. Yu (Eds.): SEWCN 2009, LNICST 42, pp. 3–12, 2010.

higher layer, more sophisticated schemes that paved the way for the transition from "weak" to "strong" secrecy, incorporating cryptography schemes [5]-[8].

Recent work, however, has re-approached the issue of physical layer-based security for wireless communication under a new light by developing the concept of Wireless Information-Theoretic Security.

1.2 Wireless Information-Theoretic Security

In [9],[10] Bloch, Barros, Rodrigues and McLaughlin suggest that the wireless communication between a transmitter and a (legitimate) receiver in the presence of a malicious user (eavesdropper) can be secure even when the SNR of the main channel is lower than the SNR of the eavesdropper. This is possible when quasi-static Rayleigh fading channels are considered, instead of the classic Gaussian scenario.

The outage probability for a given Secrecy Rate $R_s > 0$ (defined as the probability that the Secrecy Capacity will be smaller than a non-zero secrecy rate) is calculated as an expression of the average main and wiretap channel SNR, $\overline{\gamma}_M$ and $\overline{\gamma}_W$ respectively:

$$P_{out}\left(C_s < R_s\right) = P_{out}(R_s) = 1 - \frac{\overline{\gamma}_M}{\overline{\gamma}_M + 2^{R_s}\,\overline{\gamma}_W}\,e^{\left(-\frac{2^{R_s}-1}{\overline{\gamma}_M}\right)} \qquad (1)$$

The practical implementation of this information-theoretic scheme can be achieved via the use of LDPC channel coding as shown in [11],[12].

1.3 Impact of the Propagation Environment

In [9],[10] the intrinsic characteristics of the propagation environment were examined by assigning a value of n=3 to the path loss exponent [13]. Thus the Outage Probability is calculated by:

$$P_{out}\left(C_s < R_s\right) = P_{out}\left(R_s\right) = 1 - \frac{e^{\left(-\frac{2^{R_s}-1}{\overline{\gamma}_M}\right)}}{1 + 2^{R_s}\left(\dfrac{d_M}{d_W}\right)^n} \qquad (2)$$

This however does not correspond to realistic cases where the path loss exponent can assume a wide range of values [14], from n=1.8 (indoor LOS cases) up to n=3.8 and even n=4 (indoor complex NLOS topology, outdoor urban shadowed dense area). In [15], the impact of this channel-dependent variation of path loss exponent on the non-zero probability of Secrecy Capacity and the Outage Probability was examined.

In all published works so far, however, another important parameter, the Outage Secrecy Capacity, has not been properly and thoroughly investigated.

2 Outage Secrecy Capacity

Outage Secrecy Capacity is defined as the maximum secrecy rate $R_s\{\max\} = C_{out}$ such that the Outage Probability is less than a certain value, i.e. p:

$$P_{out}\left(C_{out}(p)\right) = p \tag{3}$$

2.1 The Need for a Closed-Form Expression

A closed-form expression for Outage Secrecy Capacity will provide us with knowledge of the largest secrecy rate for a given Outage Probability, intrinsic channel characteristics (path loss exponent) and average main channel SNR, namely the exact value of the secrecy rate that will serve as a threshold for Secrecy Capacity.

2.2 Closed-Form Expression via Taylor Approximation

The approximation of the exponential function via a first-order Taylor series in its generalized expression is provided by [16]:

$$e^x = \sum_{n=0}^{\infty} \frac{x^n}{n!} = 1 + x + \frac{x^2}{2!} + \frac{x^3}{3!} + \frac{x^4}{4!} + \dots \tag{4}$$

In our case, the approximation via a first-order Taylor series is achieved as such:

$$e^{\left(-\frac{2^{R_s}-1}{\overline{\gamma}_M}\right)} \approx 1 - \left(\frac{2^{R_s}-1}{\overline{\gamma}_M}\right) \tag{5}$$

Thus, the Outage Probability is provided by:

$$P_{out}\left(R_s\right) = 1 - \frac{1 - \left(\dfrac{2^{R_s}-1}{\overline{\gamma}_M}\right)}{1 + 2^{R_s}\left(\dfrac{d_M}{d_W}\right)^n} \tag{6}$$

The closed-form expression for Outage Secrecy Capacity for a given Outage Probability p is given by:

$$C_{out}(p) = \log_2\left(\frac{\overline{\gamma}_M\, p + 1}{\overline{\gamma}_W\,(1-p) + 1}\right) \tag{7}$$

$$C_{out}(p) = \log_2 \left(\frac{p + \dfrac{1}{\overline{\gamma}_M}}{\left(\dfrac{d_M}{d_W}\right)^n (1-p) + \dfrac{1}{\overline{\gamma}_M}} \right) \qquad (8)$$

Naturally, a certain error lies within this approach. Even though the parameters in Eq. 5 are such (target Secrecy Rate, average main channel SNR) that their value range allows for the approximation to take place, it is imperative to evaluate the precise error for realistic values of these parameters, that is for realistic schemes that we will be compelled to resolve in actual scenarios of information-theoretic security. In the following section, the exact calculation of this error for two different channel cases is accomplished and the findings, based on computation of Outage Probability, are discussed.

3 Error Calculation Based on Outage Probability

Two channel cases will be examined, corresponding to two very frequently met scenarios in wireless communications: (a) The pass loss exponent is assigned a value of n=2. This corresponds to the Free Space Model and describes a "good" channel case where the attenuation of the average signal strength follows the inverse-square law [17], and (b) the pass loss exponent is assigned a value of n=3.8 that stands for a "bad" channel case with heavy attenuation of the average signal strength. This corresponds to urban shadowed outdoor propagation or obstructed indoor propagation (corridors, complex topologies).

The average main channel SNR is assigned a value of 10 dB for the first case. For the n=3.8 case, the average main channel SNR is set to 0 dB.

3.1 Error Calculation for "Good" Channel Case

The following tables provide an analytical presentation of the error of our approximation for different distance ratios and for various realistic values of the target Secrecy Rate. It is interesting to note that for values of the Secrecy Rate above 3.5 bits per second, the Taylor approximation gives an Outage Probability of 1, whereas the original formula gives also very high values of Outage Probability. It is therefore a common assumption, despite some minor error deviation present (especially for the distance ratio scheme of dw=5dm) that for the given channel characteristics the Secrecy Rate should not exceed 3.5 bits per second otherwise the communication is compromised in terms of information-theoretic security.

The error is quite small for realistic values of the Secrecy Rate. For Rs< 1 bit/s, the mean error is below 1%. The curves confirm the correctness of our approach, while demonstrating that the Taylor approximation is more linear and faster than the original formula containing the exponential function.

Table 1. Comparative Calculation of Outage Probability for n=2 and avg. SNR=10 dB

Rs	dw = 5 dm		dw = 2 dm		dm = dw		dm = 2 dw	
	Exp.	Taylor	Exp.	Taylor	Exp.	Taylor	Exp.	Taylor
0,01	0,039	0,039	0,202	0,202	0,502	0,502	0,801	0,801
0,5	0,092	0,093	0,291	0,292	0,603	0,603	0,856	0,856
1	0,162	0,167	0,397	0,4	0,698	0,7	0,9	0,9
1,5	0,252	0,266	0,512	0,521	0,782	0,787	0,932	0,934
2	0,361	0,397	0,630	0,65	0,852	0,86	0,956	0,956
2,5	0,488	0,564	0,74	0,779	0,906	0,92	0,973	0,977
3	0,624	0,773	0,835	0,9	0,945	0,967	0,985	0,991
3,5	0,755	1	0,906	1	0,971	1	0,992	1
4	0,864	1	0,955	1	0,987	1	0,997	1
4,5	0,940	1	0,983	1	0,995	1	0,999	1
5	0,980	1	0,995	1	0,998	1	0,999	1

Table 2. Mean Error (%) for the Taylor Approximation for n=2 and avg. SNR=10 dB

Rs (bits/s)	dw=5dm	dw=2dm	dm=dw	dm=2dw
0< Rs ≤ 5	10,47	3,31	0,96	0,25
0< Rs ≤ 3	8,35	2,73	0,81	0,21
0< Rs ≤ 1	1,21	0,34	0,10	0,02

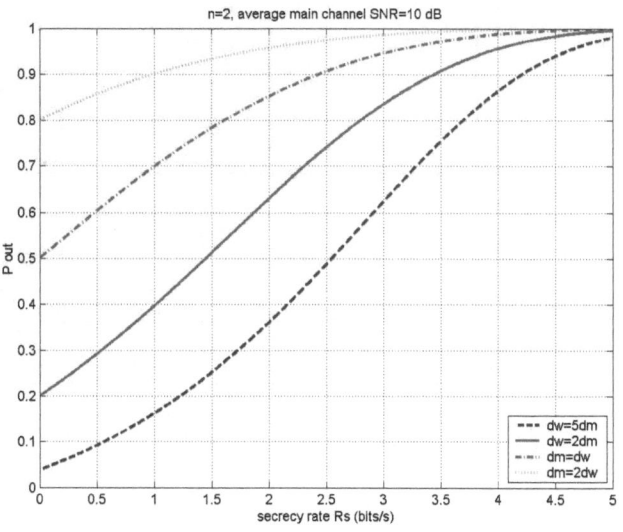

Fig. 1. Outage Probability versus Secrecy Rate for n=2 and avg. SNR=10 dB

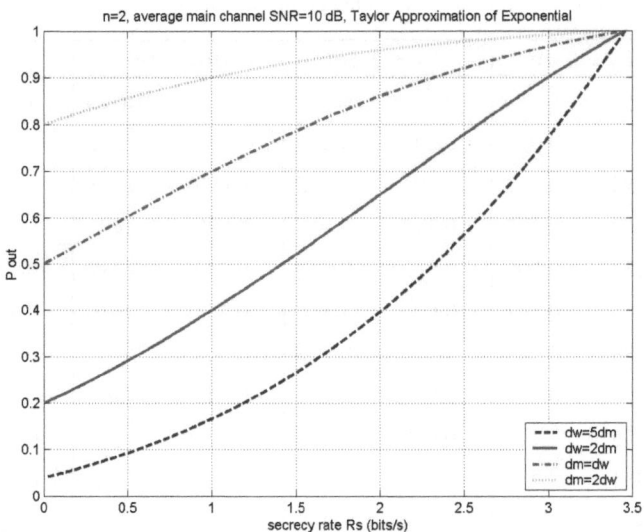

Fig. 2. Outage Probability versus Secrecy Rate for n=2 and avg. SNR=10 dB (Taylor approx.)

3.2 Error Calculation for "Bad" Channel Case

Table 3. Comparative Calculation of Outage Probability for n=3.8 and avg. SNR=0 dB

Rs	dw = 5 dm		dw = 2 dm		dm = dw		dm = 2 dw	
	Exp.	Taylor	Exp.	Taylor	Exp.	Taylor	Exp.	Taylor
0,01	0,009	0,009	0,074	0,074	0,505	0,505	0,934	0,934
0,5	0,341	0,416	0,400	0,468	0,726	0,757	0,968	0,972
1	0,634	1	0,678	1	0,877	1	0,987	1
1,5	0,840	1	0,867	1	0,958	1	0,996	1
2	0,951	1	0,961	1	0,99	1	0,999	1
2,5	0991	1	0,993	1	0,998	1	0,999	1

Table 4. Mean Error (%) for the Taylor Approximation for n=3.8 and avg. SNR=0 dB

Rs (bits/s)	dw=5dm	dw=2dm	dm=dw	dm=2dw
0< Rs ≤ 3	13,12	7,69	2,16	0,20
0< Rs ≤ 0,5	10,96	8,51	2,14	0,19
Rs=0.01	0	0	0	0

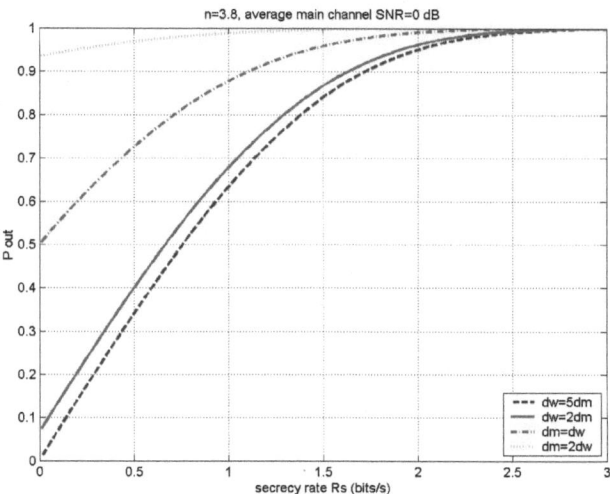

Fig. 3. Outage Probability versus Secrecy Rate for n=3.8 and avg. SNR=0 dB

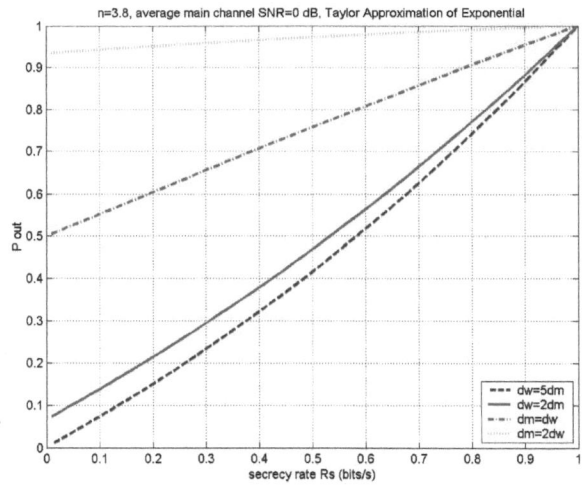

Fig. 4. Outage Probability versus Secrecy Rate for n=3.8 and avg. SNR=0 dB (Taylor approx.)

4 Behavior of the Closed-Form Expression

We shall examine, for the two aforementioned channel cases, the behavior of the Outage Secrecy Capacity that determines the threshold of secure communications for the information-theoretic scheme:

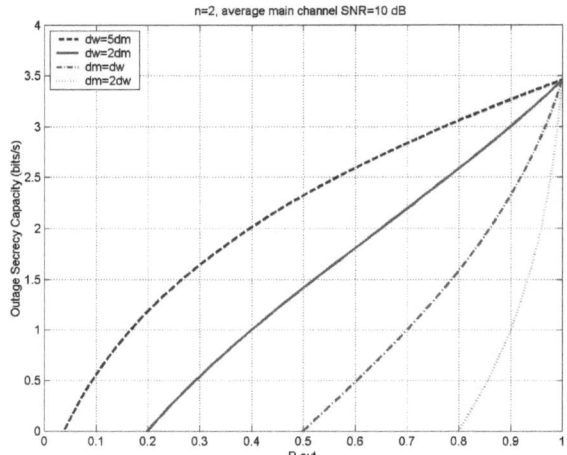

Fig. 5. Outage Secrecy Capacity for n=2, avg. SNR=10 dB and various distance ratio schemes

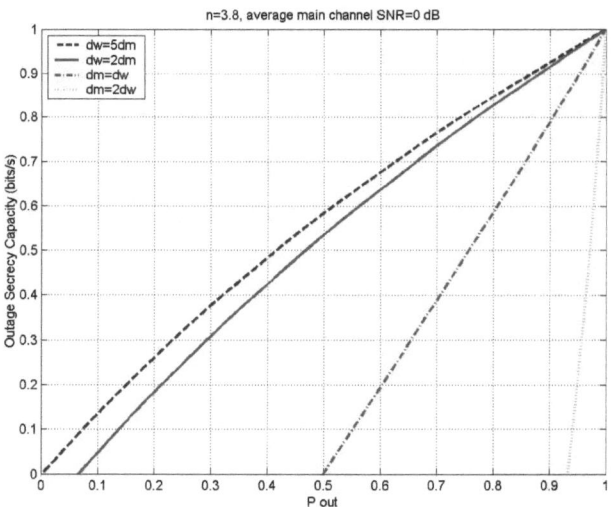

Fig. 6. Outage Secrecy Capacity for n=3.8, avg. SNR=0 dB and various distance ratio schemes

It can be observed that for n=2 and a main channel SNR=10 dB, the largest possible Secrecy Rate is smaller than 3.5 bits per second, whereas for an acceptable value of Outage Probability the largest Secrecy Rate should not exceed 1.5 bits per second. Of course the actual optimal value of the largest Secrecy Rate, namely the Outage Secrecy Capacity, depends on the specific value of the distance ratio as well, thus demonstrating the importance of the users' location.

For n=3.8 and SNR=0 dB (main channel), the largest possible Secrecy Rate is well below 1 bit per second, whereas for an acceptable value of Outage Probability, the Secrecy Rate should not exceed 0.5 bits per second, again depending on the actual user location in the topology in question.

5 Conclusions

A closed-form expression for Outage Secrecy Capacity was provided in this work, based on an approximation of the exponential function in the original formula of Outage Probability via Taylor series (first-order). The mean error was calculated for certain regions of the Secrecy Rate and the precision was evaluated as satisfactory for realistic scenarios. Therefore the closed-form expression and the subsequent curves are able to predict the behavior of the Outage Secrecy Capacity in reliable and practical manner.

The above curves provide an essentially significant standpoint for the precise estimation of the required threshold for a given Outage Probability when the propagation characteristics of the topology at hand and the location of the users therein play an important role in the reliability of information-theoretic security schemes.

Acknowledgments. The authors wish to acknowledge the support of the ICT European Research Programme and all the partners in PEACE: PDMF&C, Instituto de Telecomunicaes, FhG Fokus, University of Patras, Thales, Telefonica, CeBit.

References

1. Shannon, C.E.: Communication theory of secrecy systems. Bell Tech. J. 29, 656–715 (1949)
2. Wyner, A.D.: The wire-tap channel. Bell Tech. J. 54, 1355–1387 (1975)
3. Csiszar, I., Korner, J.: Broadcast channels with confidential messages. IEEE Trans. Inf. Th. 24(3), 339–348 (1978)
4. Leung-Yan-Cheong, S.K., Hellman, M.E.: The Gaussian wiretap channel. IEEE Trans. Inf. Th. 24(4), 451–456 (1978)
5. Maurer, U.M.: Secret key agreement by public discussion from common information. IEEE Trans. Inf. Th. 39(3), 733–742 (1993)
6. Maurer, U.M.: Information-theoretically secure secret-key agreement by NOT authenticated public discussion. In: Fumy, W. (ed.) EUROCRYPT 1997. LNCS, vol. 1233, pp. 209–225. Springer, Heidelberg (1997)
7. Maurer, U.M., Wolf, S.: Information-theoretic key agreement: from weak to strong secrecy for free. In: Preneel, B. (ed.) EUROCRYPT 2000. LNCS, vol. 1807, pp. 351–368. Springer, Heidelberg (2000)
8. Maurer, U.M., Wolf, S.: Secret-key agreement over unauthenticated public channels Part I: Definitions and a completeness result. IEEE Trans. Inf. Th. 49(4), 822–831 (2003)
9. Barros, J., Rodrigues, M.R.D.: Secrecy capacity of wireless channels. In: 2006 IEEE International Symposium on Information Theory, pp. 356–360. IEEE Press, New York (2006)
10. Bloch, M., Barros, J., Rodrigues, M.R.D., McLaughlin, S.W.: Wireless Information-Theoretic Security. IEEE Trans. Inf. Th. 54(6), 2515–2534 (2008)
11. Bloch, M., Thangaraj, A., McLaughlin, S.W., Merolla, J.M.: LDPC-based Gaussian key reconciliation. In: 2006 IEEE Information Theory Workshop, pp. 116–120. IEEE Press, New York (2006)
12. Richardson, T.J., Shokrollahi, M.A., Urbanke, R.L.: Design of capacity-approaching irregular low-density parity-check codes. IEEE Trans. Inf. Th. 47(2), 619–637 (2001)

13. Rappaport, T.: Wireless Communications: Principles and Practice. Prentice Hall, Upper Saddle River (2001)
14. Seybold, J.: Introduction to RF Propagation. Wiley Interscience, Hoboken (2005)
15. Chrysikos, T., Kotsopoulos, S.: Impact of channel-dependent variation of path loss exponent on Wireless Information-Theoretic Security. In: Wireless Telecommunications Symposium 2009, April 22-24, pp. 1–7. IEEE Press, New York (2009)
16. Abramowitz, M., Stegun, I.: Handbook of Mathematical Functions with Formulas, Graphs, and Mathematical Tables. Dover Publications, New York (1970)
17. Kotsopoulos, S., Karagiannidis, G.: Mobile Communication. Papasotiriou SA Publication, Athens (1997)

Enhanced Access Polynomial Based Self-healing Key Distribution

Ratna Dutta[1], Sourav Mukhopadhyay[2], and Tom Dowling[1]

[1] Claude Shannon Institute, Computer Science Department, NUI Maynooth, Co.
Kildare, Ireland
{rdutta,tdowling}@cs.nuim.ie
[2] School of Electronic Engineering, Dublin City University, Dublin 9, Ireland
msourav@eeng.dcu.ie

Abstract. A fundamental concern of any secure group communication
system is that of key management. Wireless environments create new key
management problems and requirements to solve these problems. One
such core requirement in these emerging networks is that of self-healing.
In systems where users can be offline and miss updates self healing allows
a user to recover lost keys and get back into the secure communication
without putting extra burden on the group manager. Clearly self healing must be only available to authorized users and this creates more
challenges in that we must ensure unauthorized or revoked users cannot,
themselves or by means of collusion, avail of self healing. To this end
we enhance the one-way key chain based self-healing key distribution of
Dutta *et al.* by introducing a collusion resistance property between the
revoked users and the newly joined users. Our scheme is based on the
concept of access polynomials. These can be loosely thought of as white
lists of authorized users as opposed to the more widely used revocation
polynomials or black lists of revoked users. We also allow each user a
pre-arranged life cycle distributed by the group manager. Our scheme
provides better efficiency in terms of storage, and the communication
and computation costs do not increase as the number of sessions grows
as compared to most current schemes. We analyze our scheme in an appropriate security model and prove that the proposed scheme is computationally secure and not only achieving forward and backward secrecy,
but also resisting collusion between the new joined users and the revoked
users. Unlike most existing schemes the new scheme allows temporary
revocation. Also unlike existing schemes, our construction does not collapse if the number of revoked users crosses a threshold value. This fea-
ture increases resilience against revocation based denial of service (DOS)
attacks and thus improves availability of communication channel.

Keywords: session key distribution, self-healing, computational security, forward and backward secrecy.

1 Introduction

In a large and dynamic group communication over an unreliable wireless network,
self-healing means that authorized users can recover the missing session keys by

Q. Gu, W. Zang, and M. Yu (Eds.): SEWCN 2009, LNICST 42, pp. 13–24, 2010.

themselves, without requesting additional transmission from the group manager. This reduces network traffic, the risk of user exposure to traffic analysis, and the work load on the group manager.

Self-healing property is being widely used for various applications. For example, mission critical applications such as in military, content sensitive internet applications such as broadcast transmission, pay per-view TV, and information distribution services.

The idea of self-healing key distribution was proposed by Staddon et al. [9]. Following it, a number of self-healing techniques have been proposed. The hash chain based schemes [3,4] are computationally secure and are highly efficient compared to the existing unconditionally secure schemes [2,6,8,12]. However, these hash chain based constructions have the fatal defect of not being collusion resistant in the sense that the collusion between new joined users and the revoked users are able to recover all the session keys which they are not entitled to.

Our contribution: In this paper, we provide a solution to the problem of resisting the collusion attack in the one-way hash chain based self-healing key distributions introduced by Dutta et al. [3,4], coupling it with the pre-arranged life cycle based approach of Tian et al. [10] that uses the same self-healing mechanism introduced in Dutta et al. [3,4]. However, we use the concept of access polynomial instead of revocation polynomial in our construction. For scalability of business it is often necessary to design more innovative and flexible business strategies in certain business models that allow contractual subscription or rental, such as subscription of mobile connection or TV channel for a pre-defined period. The subscribers are not allowed to revoke before their contract periods (life cycles) are over. Our scheme fits into such business strategies. Our construction is flexible and robust in the sense that there is no restriction on the number of revoked users, any number of users can leave/join the group and a user can join/leave as many times as she wishes. Consequently, the availability of communication channel is increased and revocation based denial of service (DOS) attacks are reduced. As compared to most existing schemes, our scheme provides better efficiency in terms of storage, and the communication and computation costs do not increase as the number of session grows, rather they increase as the number of authorized users in a session grows. While most of the existing schemes collapse when the number of revoked users crosses a threshold value, say t, our scheme is unaffected by this limitation. Moreover, if the number of authorized users remains less than t, the communication and computation cost in our scheme are significantly less than that in the existing schemes together with less storage overhead. These are the most important features of our construction. The proposed scheme is proven to be computationally secure and achieve forward and backward secrecy together with resisting collusion between the newly joined users and the revoked users. The security analysis is in an appropriate security model.

2 Preliminaries

2.1 Notational Convention

The following notations are used throughout the paper.

$\mathcal{U}$: set of all users in the networks
U_i : i-th user
GM : group manager
n : total number of users in the network
m : total number of sessions
Auth_j : the set of all authorized users in session j
F_q : a field of order q
S_i : personal secret of user U_i
SK_j : session key generated by the GM in session j
$\mathcal{B}_j$: broadcast message by the GM during session j
$Z_{i,j}$: the information learned by U_i through $\mathcal{B}_j$ and S_i

2.2 Our Security Model

We state the following definitions that are aimed to computational security for session key distribution adopting the security model of [7,9].

Definition 1 *(Session Key Distribution with privacy [9]). Let $i \in \{1, \ldots, n\}$ and $j \in \{1, \ldots, m\}$.*

1) $\mathcal{D}$ is a session key distribution with privacy if

(a) for any user U_i, the session key SK_j is efficiently determined from $\mathcal{B}_j$ and S_i.

(b) for any set $R \subseteq \mathcal{U}$ of revoked users and $U_i \notin R$, it is computationally infeasible for users in R to determine the personal key S_i.

(c) If we consider separately either the set of m broadcasts $\{\mathcal{B}_1, \ldots, \mathcal{B}_m\}$ or the set of n personal keys $\{S_1, \ldots, S_n\}$, then it is computationally infeasible for users $U_1, \ldots, U_n$ to compute session key SK_j (or other useful information) from either set. Information from both the sets is required in order to compute SK_j or any useful information.

2) $\mathcal{D}$ has revocation capability if given any $R \subseteq \mathcal{U}$ of users revoked in and before session j, the group manager GM can generate a broadcast $\mathcal{B}_j$, such that for all $U_i \notin R$, U_i can efficiently recover the session key SK_j, but the revoked users cannot. i.e. it is computationally infeasible to compute SK_j from $\mathcal{B}_j$ and $\{S_l\}_{U_l \in R}$.

3) $\mathcal{D}$ is self-healing if the following is true for any j, $1 \leq j_1 < j < j_2 \leq m$:

(a) For any user U_i who is a member in sessions j_1 and j_2, the key SK_j is efficiently determined by the set $\{Z_{i,j_1}, Z_{i,j_2}\}$.

(b) Let $1 \leq j_1 < j < j_2 \leq m$. For any disjoint subsets $L_1, L_2 \subset \mathcal{U}$, where the set L_1 is a coalition of users removed before and in session j_1 and the set L_2 is a coalition of users joined since session j_2, the set $\{Z_{l,j}\}_{U_l \in L_1, 1 \leq j \leq j_1} \cup \{Z_{l,j}\}_{U_l \in L_2, j_2 \leq j \leq m}$ cannot determine the session key SK_j, $j_1 < j < j_2$. i.e. SK_j can not be obtained by the coalition $L_1 \cup L_2$. This is collusion resistance property for self-healing.

Definition 2 *(Forward and backward secrecy [7]). Let $i \in \{1, \ldots, n\}$ and $j \in \{1, \ldots, m\}$.*

1) A key distribution scheme $\mathcal{D}$ guarantees forward secrecy if for any set $R \subseteq \mathcal{U}$ of users revoked in and before session j, it is computationally infeasible for the members in R together to get any information about SK_j, even with the knowledge of group keys $\mathsf{SK}_1, \ldots, \mathsf{SK}_{j-1}$ before session j.

2) A session key distribution $\mathcal{D}$ guarantees backward secrecy if for any set $J \subseteq \mathcal{U}$ of users joined after session j, it is computationally infeasible for the members in J together to get any information about SK_j, even with the knowledge of group keys $\mathsf{SK}_{j+1}, \ldots, \mathsf{SK}_m$ after session j.

3   Proposed Scheme

For our construction, we consider a setting in which there is a group manager (GM) and n users $\mathcal{U} = \{U_1, \ldots, U_n\}$. All operations take place in a finite field, F_q, where q is a large prime number ($q > n$). In our setting, we allow a revoked user to rejoin the group in a later session. Let $\mathcal{H} : F_q \longrightarrow F_q$ be a cryptographically secure one-way function. See [5] for a formal definition of one-way function. We use Cryptographically Secure Pseudo random bit Generators (CSPRBG) in our construction. An example of CSPRBGs include the RSA PBG [1].

3.1   Key Distribution

• *Setup*: The group manager randomly picks an initial backward key seed $S^B \in F_q$. It repeatedly applies the one-way function $\mathcal{H}$ to compute the one-way key chain of length m: $K_i^B = \mathcal{H}(K_{i-1}^B) = \mathcal{H}^{i-1}(S^B)$ for $1 \leq i \leq m$. The GM also selects at random n numbers $\alpha_1, \ldots, \alpha_n \in F_q$ and m numbers $\beta_1, \ldots, \beta_m \in F_q$ by running a CSPRBG which is cryptographically secure. The j-th session key is computed as $\mathsf{SK}_j = \beta_j + K_{m-j+1}^B$. Unlike the existing self-healing key distribution schemes, our setting allows a revoked user to rejoin the group in a later session with a new identity. However, we make the following restriction on the life cycle of each user as determined by the GM. Each user U_i is first assigned a pre-arranged life cycle (s_i, t_i), where $1 \leq s_i < t_i \leq m$, by the GM. *i.e.* U_i is involved in $k_i = t_i - s_i + 1$ many sessions and is not allowed to revoke before session t_i. However U_i may go off-line during its life cycle due to power failure. Self-healing is needed at this point. Each user U_i, for $1 \leq i \leq n$, receives its personal secret keys corresponding to the $k_i = t_i - s_i + 1$ sessions $S_i = \{\alpha_i; \beta_{s_i}, \ldots, \beta_{t_i}\}$ from the group manager via the secure communication channel between them.

• *Broadcast*: Let Auth_j be the set of all authorized (active) users in session j. In the j-th session, the group manager randomly chooses a blind value $\theta_j \in F_q$, $\theta_j \notin \{\alpha_1, \ldots, \alpha_n\}$, locates the backward key K_{m-j+1}^B in the backward key chain and computes the polynomials: $A_j(x) = 1 + (x - \theta_j) \prod_{\{l:U_l \in \mathsf{Auth}_j\}} (x - \alpha_l)$, $h_j(x) = K_{m-j+1}^B A_j(x)$. The polynomial $A_j(x)$ is called the *access polynomial* in session j. The factor $(x - \theta_j)$ is a blinding term and $\theta_j \in F_q$ is randomly selected for each session j and is different from $\alpha_1, \ldots, \alpha_n \in F_q$. The purpose of $(x - \theta_j)$ is to make $A_j(x)$ different for different session j even they contain the same α's of authorized users. Note that $A_j(\alpha_i) = 1$ for $U_i \in \mathsf{Auth}_j$. However, it is random for an unauthorized user. The group manager broadcasts the following message $\mathcal{B}_j = \{h_j(x)\}$.

- *Session Key Recovery:* When an authorized (non-revoked) user $U_i \in \mathsf{Auth}_j$ receives the j-th session key distribution message $\mathcal{B}_j$, it recovers $K^B_{m-j+1} = h_j(\alpha_i)$ as $A_j(\alpha_i) = 1$. Finally, $U_i \in \mathsf{Auth}_j$ evaluates the current session key $\mathsf{SK}_j = \beta_j + K^B_{m-j+1}$. An unauthorized user cannot construct the polynomial $A_j(x)$ as it does not know the α-values of the set of authorized users Auth_j in session j and the blind value θ_j used in session j.
- *Add Group Members:* When a new user wants to join the communication group starting from session j, the user gets in touch with the GM. The GM in turn picks an unused identity $v \in F_q$, selects a new $\alpha_v \in F_q$, assigns a life cycle (s_v, t_v) to the new user with $s_v = j$, computes the personal secret keys corresponding to $k_v = t_v - s_v + 1$ sessions $S_v = \{\alpha_v; \beta_{s_v}, \ldots, \beta_{t_v}\}$ and gives S_v to this new group member via the secure communication channel between them.

Complexity

- *Storage overhead:* Storage complexity of personal key for user U_i with life cycle (s_i, t_i) is $(t_i - s_i + 2) \log q$ bits.
- *Communication overhead:* Communication bandwidth for key management at the j-th session is $(|\mathsf{Auth}_j| + 2) \log q$ bits, where Auth_j is the set of authorized users in session j.
- *Computation overhead:* The computation cost for key management at the j-th session is $(|\mathsf{Auth}_j| + 1)$, which is the number of multiplication operations needed to find a point on a $|\mathsf{Auth}_j| + 1$-degree polynomial.

3.2 Self-healing

We now explain our self-healing mechanism for the construction. Let U_i be a group member that receives session key distribution messages $\mathcal{B}_{j_1}$ and $\mathcal{B}_{j_2}$ in sessions j_1 and j_2 respectively, where $1 \leq j_1 \leq j_2$, but not the session key distribution message $\mathcal{B}_j$ for session j, where $j_1 < j < j_2$. User U_i can still recover all the lost session keys K_j for $j_1 < j < j_2$ as desired by Definition 1 $3(a)$ using the following steps.

- U_i recovers from the broadcast message $\mathcal{B}_{j_2}$ in session j_2, the backward key $K^B_{m-j_2+1}$ and repeatedly apply the one-way function $\mathcal{H}$ on this and computes the backward keys K^B_{m-j+1} for all j, $j_1 \leq j < j_2$.
- U_i then recovers all the session keys $\mathsf{SK}_j = \beta_j + K^B_{m-j+1}$, for $j_1 \leq j \leq j_2$.

Note that a user U_i revoked in session j cannot compute the backward keys K^B_{m-j+1} for $j_1 > j$. Moreover, since a user is not allowed to revoke before the end of its life cycle, U_i revoked in j-th session means its life cycle completes at the j-th session. Consequently, U_i does not have β_{j_1} for $j_1 > j$. As a result, revoked users cannot compute the subsequent session keys SK_{j_1} for $j_1 > j$, as desired. This is forward secrecy.

Similarly, a user U_i joined in session j does not have β_{j_2} for $j_2 < j$, although it can compute the backward keys $K^B_{m-j_2+1}$ for $j_2 < j$. This forbids U_i to compute the previous session keys as desired. This is backward secrecy.

Now we will show that our construction can resist collusion required by Definition 1 3(b). Let $1 \leq j_1 < j < j_2 \leq m$. For any disjoint subsets $L_1, L_2 \subset \mathcal{U}$, let the set L_1 is a coalition of users removed before and in session j_1 and the set L_2 is a coalition of users joined from session j_2. Then no information about the session key SK_j, $j_1 < j < j_2$ can be obtained by the coalition $L_1 \cup L_2$. Our construction satisfies this property as illustrated below: Secret information held by users in $L_1 \cup L_2$ and broadcasts in all the sessions do not get any information about SK_j for $j_1 < j < j_2$. This is true because in the worst case, the coalition knows $S_i = \{\alpha_i; \beta_1, \ldots, \beta_{j_1-1}\}$ for $U_i \in L_1$, $S_i = \{\alpha_i; \beta_{j_2}, \ldots, \beta_m\}$ for $U_i \in L_2$, and $\mathcal{B}_1, \ldots, \mathcal{B}_m$. For each session j, $j_1 < j \leq j_2 - 1$, the coalition can get backward key K^B_{m-j+1} from L_2. However the session key SK_j is computed from the backward key K^B_{m-j+1} and a random number β_j. The coalition $L_1 \cup L_2$ cannot obtain the random numbers β_j for $j_1 < j < j_2$. Consequently, all the guess for SK_j with $j_1 < j < j_2$ are equi-probable.

4   Security Analysis

Theorem 3. *Our construction is secure, self-healing session key distribution scheme with privacy, revocation capability with respect to Definition 1 in our security model as described in Section 2.2 and achieves forward and backward secrecy with respect to Definition 2 in the model.*

Proof: Our goal is security against coalition of any size. We will show that our construction is computationally secure with respect to revoked users assuming the difficulty of inverting one-way function, *i.e.* for any session j it is computationally infeasible for any set of revoked users before and in session j to compute with non-negligible probability the session key SK_j, given the View consisting of personal keys of revoked users, broadcast messages before, in and after session j and session keys of revoked users before session j.

Consider a coalition R_j of users revoked in or before the j-th session. The revoked users are not entitled to know the j-th session key SK_j. We can model this coalition R_j as a polynomial-time algorithm $\mathcal{A}'$ that takes View as input and outputs its guess for SK_j. We say that $\mathcal{A}'$ is successful in breaking the construction if it has a non-negligible advantage in determining the session key SK_j. Then using $\mathcal{A}'$, we can construct a polynomial-time algorithm $\mathcal{A}$ for inverting one-way function $\mathcal{H}$ and have the following claim:

Claim: Assuming a cryptographically secure CSPRBG, $\mathcal{A}$ inverts one-way function $\mathcal{H}$ with non-negligible probability if $\mathcal{A}'$ is successful.

Proof: Given any instance $y = \mathcal{H}(x)$ of one-way function $\mathcal{H}$, $\mathcal{A}$ first generates an instance View for $\mathcal{A}'$ as follows: $\mathcal{A}$ randomly generates n distinct numbers $\alpha_1, \ldots, \alpha_n \in F_q$ and m distinct numbers $\beta_1, \ldots, \beta_m \in F_q$ by running a cryptographically secure CSPRBG and constructs the following backward key chain by repeatedly applying $\mathcal{H}$ on y: $K_1^B = y, K_2^B = \mathcal{H}(y), K_3^B = \mathcal{H}^2(y), \ldots, K_j^B = \mathcal{H}^{j-1}(y), \ldots, K_m^B = \mathcal{H}^{m-1}(y)$. $\mathcal{A}$ computes the j-th session key $\mathsf{SK}_j = \beta_j + K_{m-j+1}^B$. For $1 \leq i \leq n$, each user $U_i \in \mathcal{U}$ with life cycle, say (s_i, t_i), $1 \leq s_i < t_i \leq m$ (which is assigned to U_i by $\mathcal{A}$), receives its personal secret keys corresponding to the k_i sessions $S_i = \{\alpha_i; \beta_{s_i}, \ldots, \beta_{t_i}\} \in F_q^{k_i+1}$ from $\mathcal{A}$ via the secure communication channel between them.

Let Auth_j be the set of all authorized (active) users in session j. In the j-th session, $\mathcal{A}$ randomly chooses a blind value $\theta_j \in F_q$, $\theta_j \notin \{\alpha_1, \ldots, \alpha_n\}$ and computes the access polynomial

$$A_j(x) = 1 + (x - \theta_j) \prod_{\{l : U_l \in \mathsf{Auth}_j\}} (x - \alpha_l)$$

and the polynomial $h_j(x) = K_{m-j+1}^B A_j(x)$. For $1 \leq j \leq m$, $\mathcal{A}$ computes broadcast message $\mathcal{B}_j$ as: $\mathcal{B}_j = \{h_j(x)\}$. Then $\mathcal{A}$ sets View as

$$\mathsf{View} = \left\{ \begin{array}{l} \alpha_k \text{ for all } U_k \in R_j; \\ \mathcal{B}_j \text{ for } j = 1, \ldots, m; \\ \beta_1, \ldots, \beta_{j-1}; \\ \mathsf{SK}_1, \ldots, \mathsf{SK}_{j-1} \end{array} \right\}$$

$\mathcal{A}$ gives View to $\mathcal{A}'$, which in turn selects $X, \beta_j' \in F_q$ randomly, sets the j-th session key to be $\mathsf{SK}_j' = \beta_j' + X$ and returns SK_j' to $\mathcal{A}$. $\mathcal{A}$ checks whether $\mathsf{SK}_j' = \mathsf{SK}_j$. If not, $\mathcal{A}$ chooses a random $x' \in F_q$ and outputs x'.

Note that the access polynomial $A_j(x)$ at the j-th session is not publicly computable from the broadcast message $\mathcal{B}_j = \{h_j(x)\}$ as:

- The set of authorized users is not transmitted publicly during broadcast.
- α-values of authorized users are used in $A_j(x)$ which are parts of secret of authorized users.
- A blinding factor $(x - \theta_j)$ is used in $A_j(x)$ where $\theta_j \in F_q$ is randomly chosen for each session j and is different from α-values of users. Thus $A_j(x)$ is different for different sessions j even if the same α-values of authorized users are used.
- $A_j(\alpha_i) = 1$ for $U_i \in \mathsf{Auth}_j$ and $A_j(\alpha_i)$ is random for $U_i \notin \mathsf{Auth}_j$.
- Computing α_i for $U_i \in \mathsf{Auth}_j$ is infeasible from the set $\{\alpha_k : U_k \notin \mathsf{Auth}_j\}$ as we assume that the CSPRBG used to generate these α-values is cryptographically secure.
- an adversary or a coalition R_j of users revoked in and before session j cannot construct the polynomial $A_j(x)$ as it does not know the α-values of the authorized users Auth_j in session j and the blind value θ_j used in session j.

From View, $\mathcal{A}'$ knows only α_k for all $U_k \in R_j$, $\beta_1, \ldots, \beta_{j-1}$ and at most $j-1$ session keys $\mathsf{SK}_1, \ldots, \mathsf{SK}_{j-1}$. Consequently $\mathcal{A}'$ has knowledge of at most $j-1$ backward keys $K_m^B, \ldots, K_{m-j+2}^B$. Observe that $\mathsf{SK}_j' = \mathsf{SK}_j$ provided

(i) the guess β'_j of $\mathcal{A}'$ for β_j is correct; and
(ii) $\mathcal{A}'$ knows the backward key K^B_{m-j+1}.

The condition (i) occurs if either of the following two holds:

- $\mathcal{A}'$ is able to choose $\beta'_j \in F_q$ so that $\beta'_j = \beta_j$, the probability of which is $1/q$ (negligible for large q).
- $\mathcal{A}'$ is able to generate β_j from View. Note that from View, $\mathcal{A}'$ knows $\beta_1, \ldots, \beta_{j-1}$ $\in F_q$. Observe that $\beta_1, \ldots, \beta_{j-1}$ are generated by a cryptographically secure CSPRBG. Thus if $\mathcal{A}'$ is able to generate β_j from the known random numbers $\beta_1, \ldots, \beta_{j-1}$, then the CSPRBG is insecure, leading to a contradiction.

The condition (ii) occurs if either of the following two holds:

- $\mathcal{A}'$ is able to compute the access polynomial $A_j(x)$ (or $A_j(\alpha_k)$ for some $U_k \in R_j$) from View and consequently can recover the backward key $K^B_{m-j+1} = h_j(x)/A_j(x)$. From View, $\mathcal{A}'$ knows α_k for all $U_k \in R_j$ and with this knowledge it is computationally infeasible for $\mathcal{A}'$ (or coalition R_j) to learn α_i for $U_i \in \text{Auth}_j$ under the security of CSPRBG. Moreover, $\mathcal{A}'$ will not be able to compute $A_j(x)$ as mentioned earlier. Consequently, $\mathcal{A}'$ will not be able to recover K^B_{m-j+1} from $\mathcal{B}_j$.
- $\mathcal{A}'$ is able to choose $X \in F_q$ so that the following relations hold:

$$K^B_m = \mathcal{H}^{j-1}(X), K^B_{m-1} = \mathcal{H}^{j-2}(X), \ldots, K^B_{m-j+2} = \mathcal{H}(X)$$

This occurs with a non-negligible probability only if $\mathcal{A}$ is able to invert the one-way function $\mathcal{H}$. In that case, $\mathcal{A}$ returns $x = \mathcal{H}^{-1}(y)$.

The above arguments show that if $\mathcal{A}'$ is successful in breaking the security of our construction, then $\mathcal{A}$ is able to invert the one-way function.   □(of claim)

Hence our construction is computationally secure under the hardness of inverting one-way function and the security of the CSPRBG. This is forward secrecy. We can also prove the computational security for backward secrecy of our construction using the similar arguments as above considering a coalition of new joined users. The only difference in the proof is that this coalition of new users joined in and after session j knows all the backward keys, but they do not know $\beta_1, \ldots, \beta_{j-1}$ and consequently are unable to compute the past session keys they were unauthorized to.

We will now show that our construction satisfies all the conditions required by Definition 1.

1) (a) Session key efficiently recovered by a non-revoked user U_i is described in the third step of our construction.

(b) For any set $R_j \subseteq \mathcal{U}$ of users revoked in and before session j, and any non-revoked user $U_i \notin R_j$, we show that the coalition R_j knows nothing about the personal secret $S_i = (\alpha_i; \beta_{s_i}, \ldots, \beta_j, \ldots, \beta_{t_i})$ of U_i with life cycle (s_i, t_i), $1 \leq s_i \leq t_i \leq m$. For any session j, U_i uses α_i and β_j as its personal secret. The coalition R_j may at most learn $\beta_1, \ldots, \beta_{j-1}$ and the probability of R_j to guess β_j is negligible under the cryptographic security of CSPRBG. Similarly, it is computationally infeasible for coalition R_j to learn α_i for $U_i \in \mathsf{Auth}_j$ from the set $\{\alpha_k : U_k \in R_j\}$ under the security of CSPRBG.

(c) The session key SK_j for the j-th session is computed from two parts: backward key K^B_{m-j+1} and random number β_j. Note that β_j is part of personal key of an unauthorized user $U_i \in \mathsf{Auth}_j$ that U_i receives from GM before or when U_i joins the group and $K^B_{m-j+1} = h_j(\alpha_i)/A_j(\alpha_i)$ is recovered by U_i from the broadcast message $\mathcal{B}_j$. Note that $A_j(\alpha_i) = 1$ for $U_i \in \mathsf{Auth}_j$ and is random for $U_i \notin \mathsf{Auth}_j$. So the personal secret keys alone do not give any information about any session key. Since the initial backward seed S^B is chosen randomly, the backward key K^B_{m-j+1} and consequently the session key SK_j is random as long as $S^B, K^B_1, K^B_2, \ldots, K^B_{m-j+2}$ are not get revealed. This in turn implies that the broadcast messages alone cannot leak any information about the session keys. So it is computationally infeasible to determine $Z_{i,j}$ from only personal key S_i or broadcast message $\mathcal{B}_j$.

2) (Revocation property) Let $R_j \subseteq \mathcal{U}$ be a set of users revoked in and before session j who collude in session j. It is impossible for coalition R_j to learn the j-th session key SK_j because the knowledge of SK_j implies the knowledge of the backward key K^B_{m-j+1}, and the knowledge of the personal secret α_i, β_j of user $U_i \in \mathsf{Auth}_j$. The coalition R_j knows the set $\{\alpha_k : U_k \in R_j\}$. The coalition R_j cannot compute α_i for $U_i \in \mathsf{Auth}_j$ from the set $\{\alpha_k : U_k \in R_j\}$ by the security of CSPRBG. Moreover, $A_j(x)$ is not publicly computable as discussed earlier. This in turn makes K^B_{m-j+1} appears random to all users in R_j. Moreover the coalition knows at most $\beta_1, \ldots, \beta_{j-1}$ and guessing β_j is negligible under the security of CSPRBG. Therefore, SK_j is completely safe from R_j in computation point of view.

3) (a) (Self-healing property) As shown in Section 3.2, user U_i can efficiently recover all missed session keys.

(b) We can prove using similar arguments as the proof of claim that our construction is computationally secure for resisting coalition under the assumption that the CSPRBG is cryptographically secure. We omit the proof here due to space constraint which will be avalible in the full version of the paper.

We now show that our construction satisfies all the conditions required by Definition 2.

1) (Forward secrecy) Let $R_j \subseteq \mathcal{U}$ and all user $U_s \in R_j$ are revoked before the current session j. The coalition R_j can not get any information about the current session key SK_j even with the knowledge of group keys before session j. This is because of the fact that in order to know SK_j, any user $U_s \in R_j$ needs to know α_i for all $U_i \in \mathsf{Auth}_j$, K^B_{m-j+1} and β_j. Determining α_i for $U_i \in \mathsf{Auth}_j$ from the set $\{\alpha_k : U_k \in R_j\}$ is infeasible by the security of CSPRBG. Hence R_j is unable to compute SK_j. Besides, because of the one-way property of $\mathcal{H}$, it is computationally infeasible to compute $K^B_{j_1}$ from $K^B_{j_2}$ for $j_1 < j_2$. The users in R_j might know the sequence of backward keys $K^B_m, \ldots, K^B_{m-j+2}$, but cannot compute K^B_{m-j+1} and consequently SK_j from this sequence. Hence our

Table 1. Comparison among different self-healing key distribution schemes in j-th session ($k_i = t_i - s_i + 1$, where (s_i, t_i) is the life cycle assigned to user U_i by the GM; Auth$_j$ is the set of authorized users in the j-th session; T_j is a threshold on the number of revoked users which depend on the monotone decreasing access structure; and t is the maximum number of revoked users)

Schemes	Storage Overhead	Communication Overhead	Computation Overhead
Construction 3 of [9]	$(m-j+1)^2 \log q$	$(mt^2 + 2mt + m + t) \log q$	$2mt^2 + 3mt - t$
Scheme 3 of [7]	$2(m-j+1) \log q$	$[(m+j+1)t + (m+1)] \log q$	$mt + t + 2tj + j$
Scheme 2 of [2]	$(m-j+1) \log q$	$(2tj+j) \log q$	$2j(t^2+t)$
Construction 1 of [6]	$(m-j+1) \log q$	$(tj+j-t-1) \log q$	$2tj+j$
Construction 1 of [3]	$(m-j+2) \log q$	$(t+1) \log q$	$2t+1$
Construction 2 of [3]	$(m-j+2) \log q$	$(t+1) \log q$	$2(t^2+t)$
Construction of [4]	$(m-j+2) \log q$	$(T_j + 1) \log q$	$2(T_j^2 + T_j)$
Construction of [10]	$(2k_i + 1) \log q$	$(T_j + 1) \log q$	$2(T_j^2 + T_j)$
Our Construction	$(k_i + 1) \log q$	$(|\text{Auth}_j| + 2) \log q$	$|\text{Auth}_j| + 1$

construction is forward secure. Moreover the coalition knows at most $\beta_1, \ldots, \beta_{j-1}$ and guessing β_j is negligible under the security of CSPRBG.

2) (Backward secrecy) Let $J_j \subseteq \mathcal{U}$ and all user $U_s \in J_j$ join after the current session j. The coalition J_j can not get any information about any previous session key SK_{j_1} for $j_1 \leq j$ even with the knowledge of group keys after session j. This is because of the fact that in order to know SK_{j_1}, any user $U_s \in J_j$ requires the knowledge of β_{j_1}. Now when a new member U_v joins the group starting from session $j+1$, the GM gives U_v at most $\beta_{j+1}, \ldots, \beta_m$, together with the value α_v. Hence it is computationally infeasible for the newly joint member to trace back for previous β_{j_1} under the security of CSPRBG for $j_1 \leq j$. Consequently, our protocol is backward secure.

5   Performance Analysis

Table 1 shows comparisons of different self-healing schemes in terms of storage, communication and computation. We use the one-way key chain based approach of self-healing mechanism introduced in [3,4] which yields computationally secure and efficient scheme as no history of revoked users are sent during broadcast.

The most prominent improvement of our scheme over the previous self-healing key distributions [2,6,7,9] is that the communication complexity and computation cost in our construction does not increase as the number of session grows, but as the number of authorized users in a session grows.

As mentioned earlier, our construction is based on [3,4]. However we have the following enhancements:

(a) No forward key chain is used in our construction unlike [3,4].

(b) We make use of access polynomial instead of revocation polynomial. Access polynomial is computable only by authorized users, whereas revocation polynomial is publicly computable.

(c) Contrary to [3,4], each U_i in our construction is pre-assigned a life cycle (s_i, t_i) by the GM following the work of [10]. Thus user U_i can participate in $k_i = t_i - s_i + 1$ sessions and can not revoke before session t_i is over.

(d) In contrast to [3,4], we have been able to resist collusion attack in our construction by using pre-selected random numbers $\beta_1, \ldots, \beta_m$ (fixed) as part of users' secret keys. A user U_i with life cycle (s_i, t_i) is given only $k_i = t_i - s_i + 1$ values $\beta_{s_i}, \ldots, \beta_{t_i}$ and a value α_i as part of its secret key by the GM via a secure communication channel between them at the initial setup. As compared to [3,4], we get less storage for our scheme. The communication and computation costs at the j-th session for our scheme are linear to Auth_j, where Auth_j is the set of authorized users in session j. Our scheme has less computation and communication overhead as compared to [3] as long as $\mathsf{Auth}_j < t$ where t is a threshold on the number of revoked users in [3].

(e) The new scheme allows temporary revocation. Unlike previous self-healing key distribution schemes, revoked users may join at later sessions with new identities without violating any security and can get only the keys of the sessions it was in. Thus our scheme is more flexible as there is no restriction on the number of revoked users. Any number of users can leave/join the group and a user can join/leave as many times as it wishes. Most of the previous schemes constrained the number of revoked users to the threshold t. If more than t users are revoked, the security of the constructions cannot be guaranteed. Our scheme overcomes this limitation and thus more practical as it increases reliability of communication channel.

(f) **Denial of service attacks:** Availability is of critical business importance from an information security and business perspective. By availability we mean that a system is working and any attack that prevents the system working is known as a denial of service (DOS) attack. DOS attacks are not interested in breaking encryption or recovering keys, just in reducing availability. DOS attack scenarios are discussed in [11]. Use of the revocation polynomial in self healing systems actually facilitates a DOS attack. The attacker in this case colludes with others to increase the number of revoked users above the threshold t thus stopping the system. Using the access polynomial approach is resilient against this attack as it does not care how many users are revoked.

We adapt the similar approach as [10] to achieve resistance to collusion attacks and the ability of revoked users to rejoin the group. However, in contrast to [10], we done away with forward hash key chains. Consequently, our scheme is more efficient than [10] in terms of both storage and computation cost. Moreover, if $|\mathsf{Auth}_j| < T_j$, the communication cost in our scheme at the j-th session is less than that in [10].

6   Conclusion

We have enhanced an existing one-way key chain based self-healing key distribution by fixing the problem of collusion attack between the revoked users and the newly joined users. We have used the concept of access polynomial and assigned a pre-arranged life cycle on each user. Our scheme is robust and efficient as compared to the most previous schemes. It does not collapse as the number of revoked users exceeds a threshold value, which increases the availability of communication channel by reducing revocation based denial of service (DOS) attacks. Our scheme does not forbid revoked users from rejoining in later sessions unlike the existing self-healing key distribution schemes. This again has commercial advantages. The proposed scheme has been proven to be computationally secure and resists collusion between new joined users and revoked users together with forward and backward secrecy in an appropriate security model. Such security properties greatly increase confidence in a system.

References

1. Alexi, Chor, Goldreich, Schnorr: RSA Rabin Bits are $1/2 + 1/\text{poly}(logn)$ secure. In: Proceedings of the IEEE 25th Annual Symposium on Foundations of Computer Science, pp. 449–557 (1984)
2. Blundo, C., D'Arco, P., Santis, A., Listo, M.: Design of Self-healing Key Distribution Schemes. Design Codes and Cryptology 32, 15–44 (2004)
3. Dutta, R., Chang, E.-C., Mukhopadhyay, S.: Efficient Self-Healing Key Distributions with Revocation for Wireless Network using One Way Key Chains. In: Katz, J., Yung, M. (eds.) ACNS 2007. LNCS, vol. 4521, pp. 385–400. Springer, Heidelberg (2007)
4. Dutta, R., Mukhopadhyay, S., Das, A., Emmanuel, S.: Generalized Self-Healing Key Distribution using Vector Space Access Structure. In: Das, A., Pung, H.K., Lee, F.B.S., Wong, L.W.C. (eds.) NETWORKING 2008. LNCS, vol. 4982, pp. 612–623. Springer, Heidelberg (2008)
5. Goldreich, O.: Foundations of Cryptography: Basic Tools. Cambridge University Press, Cambridge (2001)
6. Hong, D., Kang, J.: An Efficient Key Distribution Scheme with Self-healing Property. IEEE Communication Letters 2005, 9, 759–761 (2005)
7. Liu, D., Ning, P., Sun, K.: Efficient Self-healing Key Distribution with Revocation Capability. In: Proceedings of the 10th ACM CCS 2003, pp. 27–31 (2003)
8. Saez, G.: On Threshold Self-healing Key Distribution Schemes. In: Smart, N.P. (ed.) Cryptography and Coding 2005. LNCS, vol. 3796, pp. 340–354. Springer, Heidelberg (2005)
9. Staddon, J., Miner, S., Franklin, M., Balfanz, D., Malkin, M., Dean, D.: Self-healing key distribution with Revocation. In: Proceedings of IEEE Symposium on Security and Privacy 2002, pp. 224–240 (2002)
10. Tian, B., Han, S., Dillon, T.-S., Das, S.: A Self-Healing Key Distribution Scheme Based on Vector Space Secret Sharing and One Way Hash Chains. In: Proceedings of IEEE WoWMoM 2008 (2008)
11. Tipton, H.: Official (ISC)2- Guide to The CISSP-CBK, 1st edn. Auerbach Publications (2006)
12. Zou, X.K., Dai, Y.S.: A Robust and Stateless Self-Healing Group Key Management Scheme. In: ICCT 2006, vol. 28, pp. 455–459 (2006)

Security Flaws in an Efficient Pseudo-Random Number Generator for Low-Power Environments

Pedro Peris-Lopez[1], Julio C. Hernandez-Castro[2], Juan M.E. Tapiador[3], Enrique San Millán[4], and Jan C.A. van der Lubbe[1]

[1] Department of Information and Communication, Delft University of Technology, The Netherlands
[2] School of Computing, Buckingham Building, Lion Terrace, Portsmouth PO1 3HE, United Kingdom
[3] Department of Computer Science, University of York, Heslington, York, YO10 5DD, United Kingdom
[4] Department of Electrical Engineering, University Carlos III of Madrid, 28911 Leganés, Spain
p.perislopez@tudelft.nl, julio.hernandez-castro@port.ac.uk, jet@cs.york.ac.uk, quique@ing.uc3m.es, j.c.a.vanderlubbe@tudelft.nl

Abstract. In 2004, Settharam and Rhee tackled the design of a lightweight Pseudo-Random Number Generator (PRNG) suitable for low-power environments (e.g. sensor networks, low-cost RFID tags). First, they explicitly fixed a set of requirements for this primitive. Then, they proposed a PRNG conforming to these requirements and using a free-running timer [9]. We analyze this primitive discovering important security faults. The proposed algorithm fails to pass even relatively non-stringent batteries of randomness such as ENT (i.e. a pseudorandom number sequence test program). We prove that their recommended PRNG has a very short period due to the flawed design of its core. The internal state can be easily revealed, compromising its backward and forward security. Additionally, the rekeying algorithm is defectively designed mainly related to the unpractical value proposed for this purpose.

Keywords: Sensor networks, RFID, PRNG, security, cryptanalysis.

1 Settharam and Rhee PRNG

In 2003, Rhee *et al.* designed an ultra-low power sensor networking platform, named i-Bean Network [7]. In this platform, sensors must support a Pseudo-Random Number Generator (PRNG) on-board for various purposes such as random transmissions delays or the generation of random packet sequence numbers. The use of Linear Congruential Generators (LCGs) or Lineal Feedback Shift Registers (LFSRs) could be appropriate due to their low hardware requirements (circuitry, memory and power consumption), but these generators are completely insecure. An alternative can be the use of a Cryptographically Secure Pseudorandom Number Generators (CSPRNG), but they are very exigent in terms of hardware demands, being unpractical in this kind of environments.

Q. Gu, W. Zang, and M. Yu (Eds.): SEWCN 2009, LNICST 42, pp. 25–35, 2010.
© Institute for Computer Sciences, Social-Informatics and Telecommunications Engineering 2010

Finally, standard cryptographic primitives such as a block cipher with a secret key working in counter mode or a hash function with a secret key operating in output feedback mode, can be used to build a PRNG. The problem here is again that these kind of constructions also exceed by far the capabilities of constrained devices. Motivated by this necessity, Seetharam and Rhee proposed an *ad hoc* simple PRNG based on a free-running timer [9]. This PRNG not only can be used for the i-Bean network, but it is suitable for other low-power environments. Prior to their design of the PRNG, the authors fixed a set of requirements that should be satisfied by their proposed generator, or any other design, for being useful within the i-Bean network environment:

- The generator must be efficient. Efficiency is defined in terms of resources, temporary requirements and memory. This property is translated into the following requirements: 1) It does not require any multiplication or division operations. It would be desirable if the generation could be achieved using just the logical operations; 2) The number of steps required for generating a single random number must be less than 10; 3) The amount of code memory used must be less than 50 bytes.
- The generator must produce a uniform distribution of random bytes (numbers in the interval [0,255]).
- The generator must not use specific features of any microcontroller, in this way being easily portable to other hardware platforms.

Upon establishing the above requirements, the authors proposed an 8-bit PRNG based on a free running timer (SR-PRNG in short). To obtain a fresh random number an XOR between the current value of the timer and the key is computed ($rv = tv \oplus key$). Next, the key and the timer values are updated. Specifically, one's complement of the timer becomes the new key ($key =\sim tv$) and the one's complement of the new random number becomes the next time value ($tv =\sim rv$). Additionally, the key is updated after the generation of K consecutive random numbers. To accomplish this task, the checksum of the received/transmited packets is used. The C-code of the proposed PRNG is included below. Exclusive-OR operation and one's complement are symbolized by "^" and "~" respectively.

```
/* Initialize Key to the ID of the node */
unsigned char key = nodeID;

/* Seetharam et al.'s PRNG */

unsigned char SR-PRNG()
{ unsigned char rv = 0;
  unsigned char tv = 0;

  tv  = get_timer();

  rv  = tv ^ key;
  key = ~tv;
```

```
tv = ~rv;

set_timer(tv);

return rv;
}
```

2   Statistical Analysis of SR-PRNG

To get any evidence of the security of a PRNG, it should be subjected to a variety of statistical tests designed to detect the specific characteristics expected from random sequences. In fact, there are different battery of tests for the evaluation of randomness. In 1995, Marsaglia introduced the Diehard tests which are a battery of stringent statistical tests for measuring the quality of a set of random numbers [5]. ENT is also another test battery, which includes the chi-square test [10]. A dedicated suite of randomness tests suitable for the evaluation of PRNG used in cryptography was proposed by the National Institute of Standards and Technology (NIST) in 2001 [8]. Recently, Sexton proposed a new series of tests [1], whose interpretation is similar to that of Diehard. Passing these battery tests is a necessary but not a sufficient condition for a generator to be secure. On the other hand, systematically passing the NIST and Diehard batteries provides strong evidence in favor of a good degree of output randomness.

In [9], authors used ENT for the evaluation of their PRNG. Specifically, ENT performs a variety of tests to the stream of bytes stored in a file. The entropy, chi-square test, arithmetic mean, Monte Carlo value for Pi, and serial correlation are the values computed. We have implemented the proposed PRNG in order to analyze its output.

There is a relevant point that is unclear in the original paper. Authors specified that the key is updated with the 8-bit Cyclic Redundancy Check (CRC) values of the transmitted and received packets. However, they do not describe any property of the messages transmitted and received in the network. In other words, we do not know the exact entropy introduced by this source.

Additionally, authors neither describe the rekeying algorithm in the paper nor facilitate us this information by personal communication. We do not know if, for example, the new key is simply the CRC of the messages passed to the channel, or instead the new key is the result of computing an XOR between the old key and the CRC.

Finally, the authors do not specify how frequently is necessary to apply the rekeying. This is a crucial point, as we demonstrate in Section 3.1. For evaluation purposes, the authors fixed this value to 10, but no justification for this number appears on the paper. In order to be able to analyze the output provided by this PRNG, we first describe the rekeying algorithm used in our experimentation (Python code is available in Appendix A).

Rekeying algorithm: In the initialization phase, we initialize a string variable A to a random value. Once the generator computes K random values ($rk = K$), the key is updated. We randomly changed the 3 most significant bits of the

Table 1. ENT Statistical Tests

Test	$rk = 10$	$rk = 100$
Entropy	7.707212	7.310092
Chi-square Test	477179.13 (0.01%)	477179.13 (0.01%)
Arithmetic Mean	126.1655	113.1528
Monte Carlo Value for Pi	2.940398806 (6.40%)	3.094900134 (1.49%)
Serial Correlation Test	-0.026964	-0.150160

variable A and compute its CRC. The key is finally updated by computing the XOR between the actual key and the result obtained from the CRC.

Once the rekeying algorithm is defined, we generate two files of 300MBytes, in the first of these files the rekeying is fixed to 10 ($rk = 10$) and in the second to 100 ($rk = 100$). If $rk > 100$, the results are catastrophic and are laking in interest. We analyze each of these files with the ENT battery as Seetharam and Rhee did. The results obtained are presented in Table 1.

First we focus on the results obtained when the rekeying factor is fixed to 10. There are many output values that offer a strong evidence of the non-randomness of the analyzed output:

– The arithmetic mean should be around 127.5 if the data were close to random. The obtained value points out that the output has a strong bias.
– The error in the Monte Carlo estimation of Pi value is huge.
– The serial correlation test evidences a slight dependence between each output byte and the previous output byte. This dependence is higher (in absolute value) as the rekeying value is incremented (see Section 3.1).
– The chi-square test is specially revealing: the percentage should be interpreted as the likelihood of the tested sequence coming from a uniform distribution. As the percentage obtained is inferior to 1% (around 1 in 10000), we can safely conclude that the studied sequence is almost certainly not random. Additionally, the measured chi-square statistic is astronomically larger than that expected (477179.1 against 255.3).

The results obtained with the rekeying factor fixed to 100 are indisputable:

– The density of information (entropy) is further reduced in comparison with the above case ($rk = 10$). Additionally, the value obtained is significantly far from the optimal value (8.0).
– The arithmetic mean indicates that the output has a very strong bias. In other words, the probabilities of ones and zeros are significantly different.
– The high value of the serial correlation coefficient points out the absolute necessity of a low rekeying factor. This factor would have to be fixed to an extremely low value, as we will see in Section 3.1. This fact evince that the core of the PRNG proposed by Seetharam and Rhee was not properly designed. Additionally, the highly negative correlation shows that the computation of complements in the PRNG is still easily observable in its output.

We have also analyzed the above two files with the Diehard battery. We will omit the results here due to their limited interest and just mention the main conclusions. In summary, the generator dramatically fails to pass each of the tests included in Diehard, even when the rekeying factor is fixed to an unpractically low value ($rk = 10$). The two sequences ($rk = 10$ or $rk = 100$) do not pass a single test, obtaining 0.0 or 1.0 for all the test p-values, and an overall p-value of 0. From all of the above, we can safely conclude that the analyzed sequences (and their generators) consistently failed the Diehard battery of tests at the 0.05 significance level.

3   Cryptanalysis of SR-PRNG

In this section we present the cryptanalysis of the PRNG proposed by Seetharam and Rhee. First, we show that their PRNG has an extraordinarily short period. Then, we demonstrate how the internal state of the generator can be easily disclosed. The above mentioned properties are extremely bad for, respectively, randomness and security reasons.

3.1   Period Evaluation

We show in the following that the period of this generator is extremely short. Specifically, every three executions the same value is generated again. In other words, if the PRNG is in the state $(tv[n], k[n])$, the following sequence is generated: $\{rv[n+1], rv[n+2], rv[n+3]\}$, $\{rv[n+1], rv[n+2], rv[n+3]\}$, etc. This can be generalized as: $rv[n+r] = rv[n+(r \mod 3)]$ for any r.

Theorem 1. *SR-PRNG returns to the same internal value each 3 iterations:*

$$\left.\begin{array}{l} key[n] = key[n+3*m] \\ tv[n] = tv[n+3*m] \end{array}\right\} For\ any\ integer\ m \tag{1}$$

Proof. We start observing the output and the internal state of three consecutive executions:

$$Iteration\quad n \tag{2}$$
$$rv[n] = tv[n] \oplus key[n]$$
$$key[n+1] = \sim (tv[n])$$
$$tv[n+1] = \sim (rv[n])$$

$$Iteration\quad n+1 \tag{3}$$
$$rv[n+1] = tv[n+1] \oplus key[n+1]$$
$$key[n+2] = \sim (tv[n+1])$$
$$tv[n+2] = \sim (rv[n+1])$$

$$Iteration \quad n+2 \tag{4}$$
$$rv[n+2] = tv[n+2] \oplus key[n+2]$$
$$key[n+3] = \sim (tv[n+2])$$
$$tv[n+3] = \sim (rv[n+2])$$

Employing the above 3 equations recursively, it easy to find that

$$key[n+3] = \sim (tv[n+2]) \tag{5}$$
$$= \sim (\sim (rv[n+1])) = rv[n+1]$$
$$= tv[n+1] \oplus key[n+1]$$
$$= \sim (rv[n]) \oplus \sim (tv[n])$$
$$= rv[n] \oplus tv[n]$$
$$= key[n]$$

and that

$$tv[n+3] = \sim (rv[n+2]) \tag{6}$$
$$= \sim (tv[n+2] \oplus key[n+2])$$
$$= \sim (\sim (rv[n+1]) \oplus \sim (tv[n+1]))$$
$$= \sim (rv[n+1] \oplus tv[n+1])$$
$$= \sim (key[n+1])$$
$$= \sim (\sim (tv[n]))$$
$$= tv[n] \qquad \qquad \square$$

This proves that after 3 iterations the generator returns to the same internal values. Therefore, the same sequence is generated again.

The authors proposed that the key have to be changed after the generation of K random values. If the rekeying is performed each K iterations $(rk = K)$, the Equation 1 can be rewritten as

$$\left. \begin{array}{l} key[n] = key[n+3*m] \\ tv[n] = tv[n+3*m] \end{array} \right\} \text{If } n+3*m < K \tag{7}$$

But the very short period is not mitigated by rekeying unless this is performed every 3 iterations. This value is unpractical and would overload the random number generation. The authors do not seem to be aware of this design problem, as they suggest the rekeying factor to be 10.

Rekeying is generally used in cryptography to limit the amount of data encrypted under the same key. A key exchange protocol is usually employed to negotiate the new key. In our case, the rekeying is used to update one of the internal values of the PRNG state. This process should introduce enough randomness in the new key, but the authors proposed the CRC of the transmitted/received packets, and as source of randomness it is not good enough. This choice presents three main problems: 1) The CRC is not a good source of randomness, as revealed by the statistical properties of the sequences analyzed; 2) The rekeying process increments both the computational load and the power consumption, and significantly reduces the throughput. To avoid these drawbacks, a high rekeying factor is usually used [3], exactly the opposite approach than the extremely low value suggested by the authors; 3) Initially, the key is set with the identification number of the node. In applications where this identification number might change (e.g. RFID systems), if this value is updated, the central back-end database and the reader get into a desynchronized state.

3.2   Disclosure of the Internal State

In this section we show how the internal secret state of the PRNG can be recovered under the very realistic assumption that only two consecutive outputs are eavesdropped. We demonstrate that SR-PRNG does not provide neither soft-forward nor soft-backward security.

Definition 1. *Soft-Forward security is the property that guarantees that an adversary listening the outputs provided by a PRNG (i.e. $rv[n+1]$, $rv[n+2]$) is not able to predict the next outputs (i.e. $rv[n+3]$, $rv[n+4]$, etc). In general terms,*

$$P_{Adv}(rv[n+k+m]|\{rv[n+i]\}_{i=1}^{k}) = \varepsilon \qquad (8)$$

for $m = 1, 2, \ldots$ and ε some negligible value and where P_{Adv} is the probability.

Definition 2. *Soft-Backward security is the property that guarantees that an attacker listening the outputs provided by the PRNG (i.e. $rv[n+1]$, $rv[n+2]$) is not able to determinate the previous state ($s[n+1] = (tv[n+1], key[n+1])$) after a new state ($sv[n+2] = (tv[n+2], key[n+2])$) has been reached:*

$$P_{Adv}(sv[n+1]|\overline{sv[n+2]}, rv[n+1], rv[n+2]) = \varepsilon \qquad (9)$$

where $\overline{sv[n+2]}$ means that $sv[n+2]$ state has been reached by the PRNG but its unknown to the attacker, and ε is a negligible value.

Theorem 2. *An adversary can recover the internat state of SR-PRNG after the eavesdropping of two consecutive outputs $\{rv[n], rv[n+1]\}$:*

$$\left.\begin{array}{l} tv[n] = rv[n] \oplus rv[n+1] \\ k[n] = rv[n+1] \end{array}\right\} \text{ For any integer } n \qquad (10)$$

Proof. We start observing the output and the internal state of two consecutive executions.

$$Iteration \quad n \tag{11}$$
$$rv[n] = tv[n] \oplus key[n]$$
$$key[n+1] = \sim (tv[n])$$
$$tv[n+1] = \sim (rv[n])$$
$$Iteration \quad n+1 \tag{12}$$
$$rv[n+1] = tv[n+1] \oplus key[n+1]$$
$$key[n+2] = \sim (tv[n+1])$$
$$tv[n+2] = \sim (rv[n+1])$$

Next, we show how an attacker is able to obtain the actual state of the generator. Once the state is known, future outputs can be computed, compromising the soft-forward security. Applying the above equations,

$$key[n+2] = \sim (tv[n+1]) \tag{13}$$
$$= \sim (\sim rv[n])$$
$$= rv[n]$$

$$tv[n+2] = \sim (rv[n+1]) \tag{14}$$

Finally, we demonstrate how the attacker is also able to acquire the previous state of the generator compromising the soft-backward security:

$$rv[n+1] = tv[n+1] \oplus key[n+1] \tag{15}$$
$$= \sim (rv[n]) \oplus \sim (tv[n])$$
$$= rv[n] \oplus tv[n]$$

From Equation 15,
$$tv[n] = rv[n] \oplus rv[n+1] \tag{16}$$

After the previous timer value is obtained, the last key value is easily obtained applying Equation 16:

$$k[n] = rv[n] \oplus tv[n] \tag{17}$$
$$= rv[n] \oplus rv[n] \oplus rv[n+1]$$
$$= rv[n+1]$$

Equations 16 and 17 can be used with $n = 0$, thus revealing the secret seed:

$$k[0] = rv[1] \tag{18}$$
$$tv[0] = rv[0] \oplus rv[1] \tag{19}$$

$\square$

Barak and Halevi proposed a formal model and a simple architecture for robust pseudorandom generators [2]. In this model backward/forward means that future/past output is secure. We switched this notation to be consistent with the conventional one. Three properties are demanded to these kind of architectures:

- Resilience: the output should look random to an observer with no knowledge of the internal state. This should hold even if the observer has complete control over the data used to refresh the internal state.
- Backward security: past output of the generator should look random to an observer, even if he knows the internal state at a later time.
- Forward security: future generator output should look random, even to an observer with knowledge of the current state, provided that the generator is refreshed with data with enough entropy.

As shown in Section 2, the output provided by SR-PRNG does not look like random at all. Backward security is compromised even if the internal state of the PRNG is not revealed (soft-backward security is not guaranteed). Therefore, SR-PRNG is not a robust PRNG, which might be necessary for designing protocols with forward and backward untraceability. In fact, this it is a sufficient but not necessary condition as Phan *et al.* showed in [6].

4 Conclusions

In this paper, we present the cryptanalysis of a lightweight PRNG suitable for low-power environments (e.g. sensor networks). We discover important security faults concerning both its core function and rekeying algorithm.

To design the core function, the authors limited the set of operations supported on-chip (i.e. PRNG) to bitwise operations (i.e. addition modulo 2 or one's complement). The bad property from the point of view of security is that all of these operations are triangular functions [4]. That is, the bit in position i in the output only depends on bits $j = 0, \dots , i$ of the input words, instead of all input bits. Furthermore, the composition of triangular operations always results in a triangular function. These undesirable characteristics greatly facilitates their successful analysis.

According to the rekeying algorithm two unfortunate decisions were taken. First of all, the *CRC* of the transmitted packets has a very low entropy, and additionally, these packets are really easy to alter by an active attacker. Secondly, a refreshment period of 10 iterations makes not sense; it would imply rekeying each 50 milliseconds, which computationally is too demanding.

Our next logical step is the great challenge of designing a secure lightweight PRNG under the requirements stated by Settharam and Rhee. Some requirements may be redefined because its specification needs further clarification (e.g. the definition of a single step).

References

1. David Sexton's battery (2005), http://www.geocities.com/da5id65536
2. Barak, B., Halevi, S.: A model and architecture for pseudo-random generation with applications to /dev/random. In: ACM Conference on Computer and Communications Security, pp. 203–212 (2005)

3. Bernstein, D.J.: Salsa20 specifications (2005),
   http://www.ecrypt.eu.org/stream/
4. Klimov, A., Shamir, A.: Cryptographic applications of T-functions. In: Matsui,
   M., Zuccherato, R.J. (eds.) SAC 2003. LNCS, vol. 3006, pp. 248–261. Springer,
   Heidelberg (2004)
5. Marsaglia, G.: The Marsaglia Random Number CDROM Including the DIEHARD
   Battery of Tests of Randomness (1996), http://stat.fsu.edu/pub/diehard
6. Phan, R.C.-W., Wu, J., Ouafi, K., Stinson, D.R.: Privacy Analysis of
   Forward and Backward Untraceable RFID Authentication Schemes (2008),
   http://www.cacr.math.uwaterloo.ca/~dstinson/papers/bfrfid-2.pdf
7. Rhee, S., Seetharam, D., Liu, S., Wang, N., Xiao, J.: i-Bean Network: An Ultra-Low
   Power Wireless Sensor Network. In: Proceedings of UBICOMP 2003 (2003)
8. Rukhin, A., Soto, J., Nechvatal, J., Smid, M., Barker, E., Leigh, S., Levenson,
   M., Vangel, M., Banks, D., Heckert, A., Dray, J., Vo, S.: A statistical test suite
   for random and pseudorandom number generators for cryptographic applications.
   NIST special publication 800-22 (2001), http://csrc.nist.gov/rng/
9. Seetharam, D., Rhee, S.: An Efficient Pseudo Random Number Generator for Low-
   Power Sensor Networks. In: Proceedings of LCN 2004, pp. 560–562. IEEE Computer Society, Los Alamitos (2004)
10. Walker, J.: Randomness Battery (1998), http://www.fourmilab.ch/random/

Appendix A

This is the source code of our implementation in Python.

```python
from random import *
from scipy import *

#define the CRC
from crc_algorithms import Crc
crc = Crc(width = 8, poly = 0x7, reflect_in = False,
xor_in = 0x0, reflect_out = False, xor_out = 0x0)

# - Begin Program -

f=open('output.dat', 'wb')

#Initialization

# a is used in the rekeying phase
a   = randint(0,2**8-1)

key = randint(0,2**8-1)
tv  = randint(0,2**8-1)

for sim in range(2**22):

    # Rekeying (rk =10 or 100)

    for rk in range(100):
            rv  = tv ^ key
            key = (~tv)%255
            tv  = (~rv)%255
            #store rv in a file
            rvs = "%c" % (rv)
            f.write(rvs)

    #rekeying
    b = randint(0,2**8-1)
    c = (b & 0xe0) | (a & 0x1f)
    cs = "%c" % (c)
    nrk= crc.table_driven(cs)
    key = nrk ^ key

# - End Program -
```

Session 2

Session Chair: Theofilos Chrysikos

RSSI-Based User Centric Anonymization for Location Privacy in Vehicular Networks

Yu-Chih Wei[1,2], Yi-Ming Chen[1], and Hwai-Ling Shan[2]

[1] Department of Information Management, National Central University
[2] Information & Communication Security Lab., Chunghwa Telecommunication Labs
{964403007,cym}@cc.ncu.edu.tw, shanhl@cht.com.tw

Abstract. In Vehicular Networks, for enhancing driving safety as well as supporting other applications, vehicles periodically broadcast safety messages with their precise position information to neighbors. However, these broadcast messages make it easy to track specific vehicles and will likely lead to compromise of personal privacy. Unfortunately, current location privacy enhancement methodologies in VANET, including Pseudonymization, K-anonymity, Random silent period, Mix-zones and path confusion, all suffer some shortcomings. In this paper, we propose a RSSI (Received Signal Strength Indicator)-based user centric anonymization model, which can significantly enhance the location privacy and at the same time ensure traffic safety. Simulations are performed to show the advantages of the proposed method. In comparison with traditional random silent period method, our method can increase at least 47% of anonymity in both simple and correlation tracking.

Keywords: VANET, Location Privacy, Tracking, Anonymity.

1 Introduction

Nowadays, more and more vehicles are equipped with navigation systems that can provide drivers with the directions to the destination. To fulfill their safety functions, many safety-related applications require that the vehicles broadcast their current position, speed, and direction periodically. Because these safety messages are broadcasted wirelessly in plaintext for safety applications, they are vulnerable to eavesdropping, and the location information of the vehicles can then be extracted from either position related data or identification related data. Therefore, although the broadcast safety messages could in principle improve driving safety, unauthorized parties or attackers may exploit the vulnerabilities of these VANET application systems and compromise the location privacy of the interested vehicles [1].

To solve the problem mentioned above, the authors in [2-5] proposed schemes to remove the correlation between locations and identifiers by periodically or randomly updating vehicles' pseudo identifiers. Although these methods could make vehicles unidentifiable within an anonymity set in motionless states, it can still be traced by movement tracking [6]. Furthermore, the locations visited by the vehicles can be associated with the places of interests [7] by firstly accumulating the driving paths, then by cross-referencing the accumulation results with geographical maps or other

Q. Gu, W. Zang, and M. Yu (Eds.): SEWCN 2009, LNICST 42, pp. 39–51, 2010.

location based services. Existing methodologies to enhance location privacy can be classified into several categories: k-anonymity [8-10], path confusion [11, 12], and Mix-Zones [13, 14]. These approaches are either of the type of centralized or partially centralized control and present two obvious drawbacks. One is the system bottleneck, the other is the potential privacy breach when the centralized control node is compromised [15].

In this paper, we propose a user-centric RSSI-based anonymity (R-Anonymity) model which can overcome the shortcomings of the previous approaches. In the aspect of adversary model, we focus on studying privacy protection of the vehicle operators under global passive adversary (GPA), which can locate and track any vehicle in a region-of-interest by eavesdropping (or intercepting) the broadcast message and utilize the adversarial RSUs deployed to estimate the locations of all broadcast message in the region-of-interest.

The rest of this paper is organized as follows. Section 2 describes the related work about location privacy enhancement. Section 3 depicts the proposed methodology of four R-Anonymity models and the anonymity algorithm. Section 4 describes how to evaluate the privacy enhancement of the proposed methodology. Section 5 discusses the simulation results of the proposed anonymity models and Section 6 presents our conclusions and future work.

2   Related Work

2.1   Identity-Based Anonymity

The idea of Identity-based anonymity approaches is to make vehicles not identifiable. According to Rongxing [1], there are two basic models for identity-based anonymity approaches: one is huge anonymous keys based(HAB), the other is group signature technique based (GSB).

In HAB [2], on board unit(OBU) stores a lot of anonymous keys, which are signed by CAs and used to sign safety messages. By changing the signing key constantly, it becomes harder to track the vehicles. The main advantage of HAB is its simplicity and straightforwardness. However it has some problems: one of these problems is that OBU needs a large storage space for the anonymous keys; another is that the key management will become a problem. Besides, processing a long list of certificate revocation list (CRL) will take a long time. In GSB [1, 16], the key idea is to allow a group member to sign messages anonymously on behalf of the group. The advantages of GSB are twofold: it reduces the number of anonymous keys and it has a shorter revocation list. But the verification time of safety messages will grow linearly with the number of revoked identities in the revocation list [1].

By incorporating HAB and GSB approaches, Calandriello [5] proposed a hybrid scheme. In this scheme, each vehicle is equipped with a group signing key and a group public key. A vehicle generates its own set of anonymous keys off line. These keys are signed by the vehicle's own private key, the latter is signed by CA to guarantee its validation. With this method, the anonymous keys of the vehicle can be generated on-the-fly and self-certified. In [17], Armknecht also proposed a similar mechanism called PKI+, which has the advantages of traditional HAB methods, but has a smaller size of CRL.

In summary, use of above methods can provide the confidentiality and integrity of messages, but as long as the vehicles need to broadcast messages periodically for the safety sake, they are still identifiable by movement tracking and profile identification of vehicle information. Hence, HAB and GSB are still unable to completely ensure the unlinkability of vehicle movement [9, 18].

2.2 Location and Attributes Based Anonymity

Full Trust Centralized Third Party

In [9], by use of a centralized third party, Shin proposed a profile based k-anonymization model to guarantee anonymity even when profiles of mobile users are known to an adversary. Vehicle privacy can be achieved by a sufficiently large k-anonymous dataset. However, the fully centralized trusted third party could become a system bottleneck or the single point of failure. Furthermore, this may lead to a serious privacy threat when the third party is attacked [15].

Path confusion uses a time-to-confusion approach to enhance location privacy in location dataset. Hoh [11] proposed an uncertainty-aware path cloaking algorithm to hide location samples in a dataset. As processing delay is a significant impediment in this method, Meyerowitz [12] tries to resolve it by predicting which intersection the users will pass through, and proactively retrieve data of the user's locations.

Partial Trust Centralized Third Party

In partial trust centralized model, vehicles could be in cooperation with road side unit (RSU) to provide location privacy without a centralized service provider. Freudiger [14] proposed a mix-zones model to enhance the location privacy. All legitimate vehicles within a mix-zone obtain a symmetric key from the RSU of that mix-zone. When a vehicle enters a RSU's mix-zone and stays within it, it uses the key provided by the RSU to encrypt all messages. However, in this scheme, there is a traffic safety concern: when vehicles come nearby a mix-zone but not yet enter the zone, it will not be able to detect its potential collision with other vehicles for lacking the encryption key of RSU.

User Centric

With user centric anonymity method, vehicles in the vehicular network are able to independently determine when and where to protect their privacy. In [19], Tang proposed a scheme called P-SRLD, which determines the relative locations of surrounding wireless connected vehicles and uses cryptographic keys to authenticate and protect driver's privacy. However, without GPS or accurate position information, his scheme does not seem to be applicable to all kinds of safety applications.

In [20], the authors propose a random silent period and swap identifiers for the two nearest vehicles. In addition to random silent period, Sampigethaya [7, 21] also proposes a group communication scheme in VANET. The scheme in [15] can also enhance privacy by using group communication. In this scheme, when a node wants to query the location-based database server through a base station, it will give a list of candidate queries that includes the actual ones and some false ones. The advantages of this method are flexibility and the property of being distributive, as in a peer-to-peer system, there is no need to have a trusted third party to store the vehicles' information. However, this method also has some drawbacks. For example, there are too many broadcast messages over the same channel which will likely cause communication jam. A more serious drawback is that keeping silent in a long time will bring about traffic mishap, especially in highways.

In next section, we propose a RSSI based anonymity method, which cannot only mitigate unauthorized location tracking but also take the vehicle safety into account.

3 Methodology

3.1 Assumptions

In order to make the proposed method work properly, we make the following assumptions.

- All of the vehicles are equipped with a GPS navigation system.
- A public key infrastructure is available in the vehicular network.
- Pseudonyms have a short validity period and cannot be reused.
- Vehicles periodically broadcast their positions, velocities and directions to the network for the sake of safety and they record these data in their individual Event Data Recorder (EDR).
- There exist fully trustworthy third parties, which conform to privacy policies and keep track of the mapping between the pseudonyms and the corresponding driver's real identity.

3.2 R-Anonymity Model

The basic idea of R-Anonymity model is to use a distance metric method to preserve the location privacy while maintaining the traffic safety. The privacy is preserved by selectively disturbing the real values (based on the risk level) of vehicles positions, directions, and velocities. These disturbed data will prevent the adversary from precisely tracking a targeted vehicle.

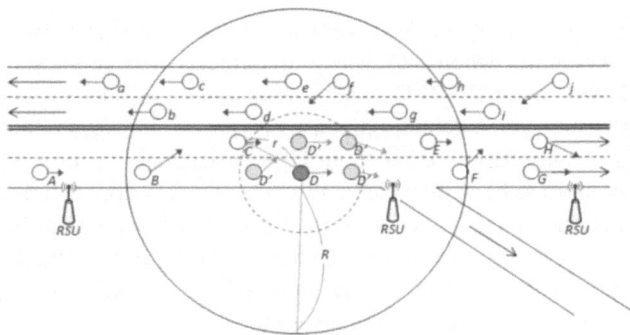

Fig. 1. Illustration of R-Anonymity model in pseudo position, direction and velocity

Take Fig. 1. as an example, the wireless transmission diameter of vehicle D is R and the vehicle C is the nearest vehicle to D with distance r. The proposed pseudo position D' will be broadcasted instead of D, where D' is within the circle centered at D with diameter r. The value of D' is chosen in a way that the pseudo position will not affect the traffic safety. It is clear that the longer the distance of the nearest vehicle, the larger range of variation of pseudo positions we could use. Consequently, as we make it more difficult for the adversary to track the correct position, we can increase the location privacy of vehicle D.

Radio Signal Strength Indication (RSSI)
The method of received signal strength indication (RSSI) is most widely known for providing a low-cost estimation of distance between vehicles [22-24]. The benefit of RSSI method is that its implementation does not require any specialized hardware. It has been used by many user location tracking applications [25-27] with an estimation error as low as 4.1m [28]. As we need a parameter to be used as a seed and a simple and fast equation to guarantee the anonymity of VANET, therefore in this paper, we leverage the RSSI to propose a user centric anonymity model, which we call R-Anonymity. Because the vehicles are in the driving mode, their moving direction, velocity and acceleration vary, hence the strengths of received signals of the vehicle are changing constantly.

$$RSSI_{ij} = -(10n \log_{10} d_{ij} + A) \tag{1}$$

In the Equation (1), $RSSI_{ij}$ indicates the radio strength from node i to node j, where i is the node of transmitter, j is the receiver, n is the path loss exponent depending on each network characteristics, d_{ij} is the distance from i to j, and A is the received signal strength at 1 meter distance. For different vendor, the range of value of RSSI is defined differently. For example, Cisco uses a range of 0~100 in their devices, while Atheros-based chipsets use a range of 0~60. In this paper, we will normalize the RSSI indicators to be from 0 to 1.

As $RSSI_{max}$ can be used to represent the distance between the tracked vehicle and its nearest vehicle, we use it to compute a pseudo value to be used in R-Anonymity model. That is, the computation of pseudo value depends on the value of $RSSI_{max}$. The radio strength is reversely proportional to the distance difference between the real value and the pseudo value, which means that when the radio strength is low, there is no vehicle close to the tracked vehicle. Hence we can generate the pseudo value with larger difference to real position to keep this vehicle hard to be tracked.

$$R_d = X \times 10^{-(RSSI_{max}+A)/10n} / C_R \quad , -1 \leq X \leq 1 \tag{2}$$

In Equation (2), R_d is the tolerable distance ratio with respect to current $RSSI_{max}$ value, X is a random value between -1 to 1, and C_R is the maximum transmission range of a VANET device. With the computation of R_d, we can provide four kinds of privacy enhancement: pseudo position, pseudo direction, pseudo velocity and pseudo random silent period, named respectively R-Pseudo position, R-Pseudo direction, R-Pseudo velocity and R-Random silent period.

R-Pseudo Position
Equation (3) shows how to generate a pseudo vehicle position value to avoid being tracked by adversaries. Here (x_r, y_r) represents the original position of the tracked vehicle, d is a constant (not greater than transmission radius) chosen to generate acceptable pseudo position.

$$(x_p, y_p) = (x_r + d \times R_d , y_r + d \times R_d) \tag{3}$$

R-Pseudo direction
In addition to vehicle's position, vehicle's driving direction should also be taken into consideration in order to reduce the risk of being tracked. This is because, if we just generate pseudo positions, adversaries can still make use of vehicle driving direction to filter the candidate vehicles and increase the probability of successful tracking.

$$D_p = D_r + C_d \times R_d \tag{4}$$

Equation (4) shows how to generate R-Pseudo direction, where D_r is the original drive direction of the node, C_d is the constant of maximum differential direction, and C_d is 30 degrees for change lane.

R-Pseudo velocity
For the similar reason as in above, vehicle velocity is also an important attribute to be considered. Equation (5) shows how to generate pseudo velocity to confuse the adversaries.

$$V_p = V_r + C_v \times R_d \tag{5}$$

In Equation (5), V_r is the original driving velocity of the tracked vehicle, and C_v is the maximum allowable velocity difference between real and pseudo velocity. In this paper, C_v is less than 5 m/sec (18 km/hr).

R-Random silent period
By altering broadcast interval, random silent period (RSP) becomes a useful method to enhance location privacy[6, 7, 20]. However, when the vehicles are close to each other, indiscriminately following random silent period rules may result in vehicle collision. To address this issue, we propose a RSSI based random silent period method. In this method, the closer distance between the vehicles, the shorter the random silent period.

$$SP_p = SP_{min} + (SP_{max} - SP_{min}) \times R_d \tag{6}$$

In Equation (6), the SP_{min} is the minimum silent period, and SP_{max} is the maximum silent period.

In order to keep safety, when a vehicle detects that the distance of the nearest vehicle is less than C_n, the broadcast messages of this vehicle will not apply the R-Anonymity model, so as to avoid possible vehicle collisions. In addition, after we apply the R-Anonymity model to obtain pseudo position, the new position might be out of the road. In order to solve this problem, we use map matching [29] to verify the pseudo position value. If the coordinates of the vehicle indicate that the pseudo position is outside of the road, the algorithm will regenerate new one and verify again, until the coordinates fall within the road.

4 Evaluation

In order to evaluate the location privacy and compare our model with other models, we first define a reachable area, A_r, to represent the next possible positions of a vehicle based on current moving directions, velocity and silent period. Then we use both simple and correlation tracking methods [7] to measure the anonymity of the tracked vehicles. The anonymity of a vehicle is determined by its anonymity set, where the latter is defined as vehicles locating within the reachable area of the tracked vehicles, but is indistinguishable from each other.

4.1 Tracking of Vehicles

Simple Tracking
With the simple tracking, it is assumed that each vehicle in the anonymity set has the same probability to be tracked vehicle. The on-coming vehicle in the reachable area is filtered according to the driving direction of vehicle. Fig.2 (a) is an example of simple tracking, in which we find vehicles $\{f, g, E, F, I\}$ are in the reachable area. But as vehicles $\{f, g\}$ are on-coming vehicles, they are filtered out of the anonymity set. Therefore, the result of the anonymity set by simple tracking is $\{E, F, I\}$.

Equations (7) are the tracking criteria for simple tracking, where V_{min} (V_{max}) is the minimum (maximum) velocity of this vehicle; SP_{min} (SP_{max}) is the minimum (maximum) silent period. Let (x_i, y_i) be the location of tracked vehicle; (x_j, y_j) be a vehicle in the transmission area. Equation (7a) is use to determine whether vehicle j is within the transmission area of vehicle i.; Equation (7b) is used to determine whether vehicle i and j are in the same driving direction.

$$\begin{cases} V_{min} \times SP_{min} \leq \sqrt{(x_i - x_j)^2 + (y_i - y_j)^2} \leq V_{max} \times SP_{max} \quad \forall j \in \{A_t\} & \text{(7a)} \\ \left| \arctan2(^{(x-x_j)}/_{(y-y_j)}) \right| \leq \pi/2 & \text{(7b)} \end{cases}$$

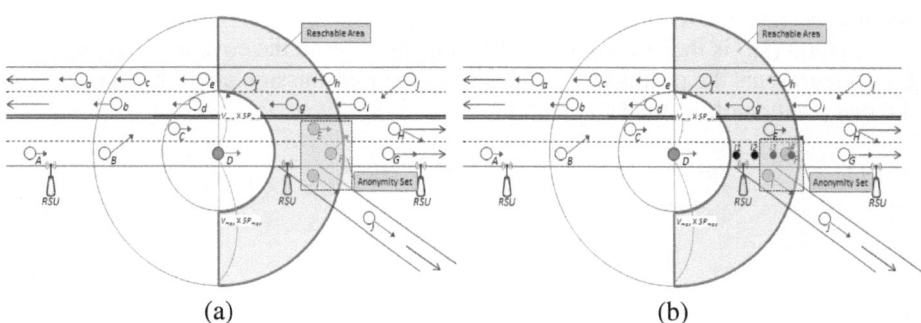

(a)   (b)

Fig. 2. (a) Simple Tracking of vehicles (b) Correlation Tracking of vehicles

Correlation Tracking
With the correlation tracking, the adversary will estimate locations of the tracked vehicle in the reachable area constantly, and then select the closest vehicle in the reachable area. The number of vehicles selected by the correlation tracking is less or equal to that selected by simple tracking. Fig.2 (b) is an example of correlation tracking, by the estimated location of tracked vehicle D in different time points from l_1 to l_4. As E is not the nearest vehicle to l_1, $l2$, $l3$, $l4$, compared with I and F, vehicle E will be filtered out, and the result of anonymity set found by correlation tracking is $\{F, I\}$.

Equations (8) are the tracking criteria for correlation tracking, where (x_t, y_t) is the estimate positions in time t, time t is between SP_{min} and SP_{max}, V_s is the vehicle velocity between V_{min} and V_{max}, θ is driving direction of the tracked vehicle, and l_{est} is the nearest vehicle in the reachable area A_r comparing with estimation positions.

$$\begin{cases} (x_t, y_t) = (x + t \times V_s \times \cos\theta, \ y + t \times V_s \times \sin\theta) \\ l_{est} = Min \ (\sqrt{(x_t - x_j)^2 + (y_t - y_j)^2}) \quad \forall j \in \{A_r\} \\ SP_{min} \leq t \leq SP_{max} \\ V_{min} \leq V_s \leq V_{max} \end{cases} \tag{8}$$

4.2 Evaluation of Anonymity

Entropy of the vehicle distribution in the anonymity set is the most popular method to evaluate the level of anonymity [7, 14]. Equation (9) is used to measure the entropy. It is assumed that the elements of the anonymity set have a uniform distribution [30]. In Equation (9), P_i is the probability of vehicle i to be selected to track, S_A is the anonymity set of vehicles, and $|S_A|$ is the number of vehicles in the anonymity set. When $H(p)=0$, it means that the tracked vehicle could not provide any anonymity set and it is easy to be tracked. The higher the value of $H(p)$, the higher the location anonymity.

$$H(p) = -\sum_{i=1}^{|S_A|} P_i \log_2 P_i, \ \sum_{i=1}^{|S_A|} P_i = 1 \tag{9}$$

For a given time t, if there is no vehicle found during t+ SP_{min} and t+ SP_{max}, it means that the tracking fails under simple or correlation tracking. This emptiness of the anonymity set makes the entropy value to become infinite. In order to resolve the infinite entropy problem, when the size of anonymity set is zero under both simple and correlation tracking, we change from the anonymity set of reachable area to the anonymity set of coverage area A_c.

Equation (10) is the criteria to get the anonymity set in the coverage area A_c, where (x_j, y_j) represents the positions of vehicles j in the transmission area (A_t), and (x_r, y_r) is the real position of the tracked vehicle.

$$\begin{cases} \sqrt{(x-x_j)^2 + (y-y_j)^2} \leq \sqrt{(x-x_r)^2 + (y-y_r)^2} \\ \left| \arctan2(^{(x-x_j)}/_{(y-y_j)}) \right| \leq {\pi}/{2} \end{cases} \quad \forall j \in \{A_t\} \tag{10}$$

$$H(p) = \log_2(|S_A|) \tag{11}$$

The number of vehicles in the coverage area of a specific vehicle is always greater or equal to 1 (this vehicle itself). Therefore, new entropy value of anonymity set (see Equation (11)) will never be infinite.

4.3 Simulation

In order to evaluate the RSSI based location privacy method, we simulate the vehicular network with two types of topology: grid and cobweb. Grid topology is a 2x2 grid, whose total length is 2196 meters with 9 intersections. Cobweb topology is a 2-circle cobweb topology with the total length of 7566 meters and 27 intersections.

We use the SUMO [31] to generate the topology of maps and use TraNS [32] to leverage the NS2 to simulate the movement of vehicles. In order to observe the effectiveness of different density of vehicles on the roads to our method, for each topology, the number of vehicles entering the topology is from 50 to 700 nodes per seconds. After NS2 finished, we import the output file of NS2 trace file to Matlab and compute the entropy in different fixed/random silent period and R-random silent period.

5 Simulation Result

In our experiments, we simulate both grid and cobweb topologies with two lanes and two drive directions. It is assumed that the maximum message transmission radius is 100 m, and the simulation is done with the range of fixed silent period from 300 to 900 ms and with random/R-random silent period from 1 to 3 seconds.

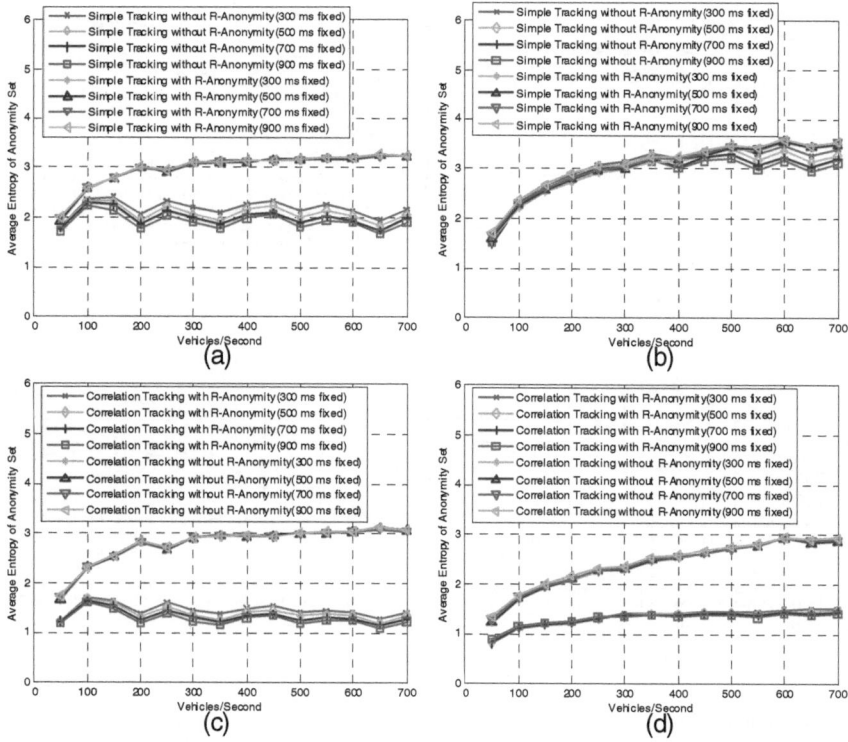

Fig. 3. Vehicle tracking to fixed silent period from 300 ms to 900 ms (a) Grid topology with simple tracking (b) Cobweb topology with simple tracking (c) Grid topology with correlation tracking (d) Cobweb topology with correlation tracking

The experimental results of fixed silent period are shown in Fig.3. The level of anonymity increases with the number of vehicles per second in the road under simple tracking. After R-Pseudo is applied in position, direction and velocity to produce interference in the simple tracking in grid topology (shown in Fig.3 (a)) and cobweb topology (shown in Fig.3 (b)), we observe that the resulting entropies in cobweb topology are higher than those in the grid topology. This is because that the number of intersections of cobweb topology is greater than that of the grid topology, which increases the difficulties of simple tracking. The entropy in the cobweb topology is relatively high so that the improvement of location privacy by R-pseudo position, direction and velocity seems less obvious.

Fig.3 (c) and (d) show the correlation tracking results of fixed silent period. From these two figures, we can find that the differences of the result in both two types of topologies are not significant in correlation tracking. It follows that the correlation tracking has higher tracking capability than simple tracking. The experimental results in both topologies with correlation tracking show that by adopting R-Pseudo method the entropy can be increased substantially.

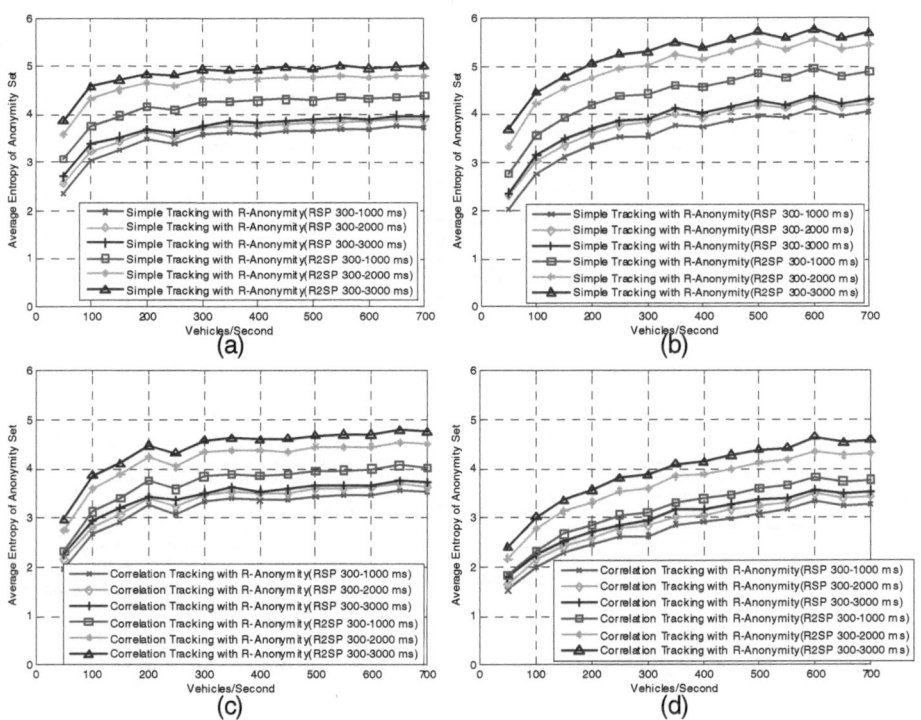

Fig. 4. Vehicle tracking to random and R-random silent period from 300ms to 3000ms (a) Grid topology with simple tracking (b) Cobweb topology with simple tracking (c) Grid topology with correlation tracking (d) Cobweb topology with correlation tracking

In regard to simple tracking and correlation tracking, Fig.4. shows that after applying R-Pseudo position, direction and velocity methods and pseudo random silent period, we can improve the anonymity under various conditions. In Fig.4 (c) and (d), the average entropy under correlation tracking is 3.46. Compared with Fig.3 (c) and (d), the overall difference of entropy between our R-pseudo method and pure correlation tracking method in Fig.4 (c) and (d) is greater than 1.91, which is quite significant. We may, therefore, conclude that the vehicular network applying R-Anonymity model can greatly enhance the location privacy and prevent vehicle from being tracked in this simulation.

6 Conclusion and Future Work

In this paper, an R-Anonymity model is proposed to cope with location privacy threats in vehicular networks. It is user centric, namely it is without the need of a centralized trusted party. Furthermore, it is simple to implement because it is easy to sample RSSI from the air. Applying R-Anonymity in vehicular networks can actually provide sufficient unlinkability to protect location privacy.

We simulate our R-Anonymity model with two kinds of topology and fourteen sets of network parameters. Simulations manifest that R-Anonymity model can obtain better results than traditional fixed and random silent periods. The results also show that by merely applying either one of R-Pseudo position, velocity, direction or R-Random silent period, the entropy value becomes relatively low, which means that the vehicles are easy to be tracked in some cases. By combining all of the above, we can provide better location privacy than the traditional methods do.

In the future, we intend to evaluate the proposed model by simulation based on the real map integrated with traffic signs. Another interesting future research direction is to address the relation between location privacy and road traffic safety by measuring the RSSI value in various conditions.

References

1. Lu, R., Lin, X., Zhu, H., Ho, P., Shen, X.: ECPP: Efficient Conditional Privacy Preservation Protocol for Secure Vehicular Communications. In: INFOCOM 2008. The 27th Conference on Computer Communications, pp. 1229–1237. IEEE, Los Alamitos (2008)
2. Raya, M., Hubaux, J.-P.: Securing vehicular ad hoc networks. Journal of Computer Security 15, 39–68 (2007)
3. Dötzer, F.: Privacy Issues in Vehicular Ad Hoc Networks. Privacy Enhancing Technologies, pp. 197–209 (2006)
4. Gerlach, M., Guttler, F.: Privacy in VANETs using Changing Pseudonyms-Ideal and Real. In: IEEE 65th Vehicular Technology Conference, 2007. VTC 2007, pp. 2521–2525 (Spring 2007)
5. Calandriello, G., Hubaux, J., Lioy, A.: Efficient and robust pseudonymous authentication in VANET. In: Proceedings of the fourth ACM international workshop on Vehicular ad hoc networks, pp. 19–28 (2007)

6. Huang, L., Matsuura, K., Yamane, H., Sezaki, K.: Towards modeling wireless location privacy. In: Danezis, G., Martin, D. (eds.) PET 2005. LNCS, vol. 3856, pp. 59–77. Springer, Heidelberg (2006)
7. Sampigethaya, K., Li, M.Y., Huang, L.P., Poovendran, R.: AMOEBA: Robust location privacy scheme for VANET. IEEE Journal on Selected Areas in Communications 25, 1569–1589 (2007)
8. Gruteser, M., Grunwald, D.: Anonymous Usage of Location-Based Services Through Spatial and Temporal Cloaking. In: Proceedings of the 1st international conference on Mobile systems, applications and services, pp. 31–42 (2003)
9. Shin, H., Atluri, V., Vaidya, J.: A Profile Anonymization Model for Privacy in a Personalized Location Based Service Environment. In: 9th International Conference on Mobile Data Management, 2008. MDM 2008, pp. 73–80 (2008)
10. Gedik, B., Liu, L.: A Customizable k-Anonymity Model for Protecting Location Privacy. Georgia Institute of Technology (2004)
11. Hoh, B., Gruteser, M., Xiong, H., Alrabady, A.: Preserving privacy in GPS traces via uncertainty-aware path cloaking. In: Proceedings of the 14th ACM conference on Computer and communications security, pp. 161–171 (2007)
12. Meyerowitz, J.T., Choudhury, R.R.: Realtime location privacy via mobility prediction: creating confusion at crossroads. In: Proceedings of the 10th workshop on Mobile Computing Systems and Applications. ACM, Santa Cruz (2009)
13. Beresford, A., Stajano, F.: Mix zones: User privacy in location-aware services. In: Proceedings of First IEEE International Workshop on Pervasive Computing and Communication Security, PerSec (2004)
14. Freudiger, J., Raya, M., Félegyházi, M., Papadimitratos, P., Hubaux, J.P.: Mix-Zones for Location Privacy in Vehicular Networks (2007)
15. Chow, C., Mokbel, M., Liu, X.: A peer-to-peer spatial cloaking algorithm for anonymous location-based service. In: Proceedings of the 14th annual ACM international symposium on Advances in geographic information systems, pp. 171–178 (2006)
16. Lin, X., Sun, X., Ho, P.H., Shen, X.: GSIS: A Secure and Privacy-Preserving Protocol for Vehicular Communications. IEEE Transactions on Vehicular Technology 56, 3442–3456 (2007)
17. Armknecht, F., Festag, A., Westhoff, D., Zeng, K.: Cross-layer privacy enhancement and non-repudiation in vehicular communication. In: 4th Workshop on Mobile Ad-Hoc Networks (WMAN) (March 2007)
18. Huang, L., Matsuura, K., Yamane, H., Sezaki, K., Japan, N.R.C., Tokyo, J.: Enhancing wireless location privacy using silent period. In: 2005 IEEE Conference on Wireless Communications and Networking, vol. 2 (2005)
19. Tang, L., Hong, X., Bradford, P.G.: Privacy-preserving secure relative localization in vehicular networks. Security and Communication Networks 1, 195–204 (2008)
20. Li, M., Sampigethaya, K., Huang, L., Poovendran, R.: Swing & swap: user-centric approaches towards maximizing location privacy. In: Proceedings of the 5th ACM workshop on Privacy in electronic society, pp. 19–28 (2006)
21. Sampigethaya, K., Huang, L., Li, M., Poovendran, R., Matsuura, K., Sezaki, K.: CARAVAN: providing location privacy for VANET. In: Proceedings of Embedded Security in Cars, ESCAR (2005)
22. Choi, W., Nam, J., Choi, S.: Hop State Prediction Method Using Distance Differential of RSSI on VANET, vol. 1 (2008)

23. Bouassida, M., Guette, G., Shawky, M., Ducourthial, B.: Sybil Nodes Detection Based on Received Signal Strength Variations within VANET. International Journal of Network Security 9, 12 (2009)
24. Bin, X., Bo, Y., Chuanshan, G.: Detection and localization of sybil nodes in VANETs. In: Proceedings of the 2006 workshop on Dependability issues in wireless ad hoc networks and sensor networks. ACM, Los Angeles (2006)
25. Lau, E.E.L., Chung, W.Y.: Enhanced RSSI-Based Real-Time User Location Tracking System for Indoor and Outdoor Environments, pp. 1213–1218 (2007)
26. Choi, W.S., Nam, J.W., Choi, S.G.: Hop State Prediction Method Using Distance Differential of RSSI on VANET, vol. 1 (2008)
27. Saxena, M., Gupta, P., Jain, B.N.: Experimental analysis of RSSI-based location estimation in wireless sensor networks, pp. 503–510 (2008)
28. Whitehouse, K., Karlof, C., Culler, D.: A practical evaluation of radio signal strength for ranging-based localization (2007)
29. Jagadeesh, G.R., Srikanthan, T., Zhang, X.D.: A Map Matching Method for GPS Based Real-Time Vehicle Location. Journal of Navigation 57, 429–440 (2004)
30. Serjantov, A., Danezis, G.: Towards an information theoretic metric for anonymity. In: Dingledine, R., Syverson, P.F. (eds.) PET 2002. LNCS, vol. 2482, pp. 41–53. Springer, Heidelberg (2003)
31. SUMO: Simulation of Urban Mobility, http://sumo.sourceforge.net/
32. Trans: Traffic and Network Simulation Environment, http://trans.epfl.ch/

Integrating Wireless Sensor Networks with Computational Grids

Nikolaos Preve

School of Electrical and Computer Engineering, National Technical University of Athens,
Heroon Polytechneiou 9, Zographou 15773, Greece
nikpreb@mail.ntua.gr

Abstract. Wireless sensor networks (WSNs) have been greatly developed and emerged their significance in a wide range of important applications such as acquisition and process information from the physical world. The evolvement of Grid computing has been based on coordination of distributed and shared resources. A Sensor Grid network can integrate these two leading technologies enabling real-time sensor data collection, the sharing of computational and storage grid resources for sensor data processing and management. Several issues have occurred from this integration which dispute the modern design of sensor grids. In order to address these issues, in this paper we propose a sensor grid architecture supporting it by a testbed which focuses on the design issues and on the improvement of our sensor grid architecture design.

Keywords: Wireless Sensor Networks, Computational Grids, Sensor Grid.

1 Introduction

Recent advances in microelectromechanical systems (MEMS), digital communications, wireless sensor networks (WSNs), distributed systems and low-power embedded processing, have led us to a new computing environment by the combination of informatics with the physical world [1]. The development of wireless sensor networks was motivated by the users' needs in order to interact with the physical world.

Nowadays, wireless sensor networks contain small, low-cost and low-power sensor nodes "motes" [2]. Also, wireless sensor networks contain sensor devices or instruments with sensing and data processing capabilities. The obvious disadvantages from the usage of sensor devices concern the limited communication bandwidth, the sensing and processing power which are also limited.

The continuous development of Grid computing led us to distributed infrastructures with shared resources. A computational grid can successfully provide a secure and convenient sharing of resources such as process power through multiple CPUs, memory, storage, content and databases in order to meet the computational requirements of users' applications. The extension of sensor grids to the grid computing was a compulsory requirement in order to overcome the limitations of the wireless sensor networks. Combining a wireless sensor network with a wired grid infrastructure we have a Sensor Grid network which is more scalable, robust and efficient than each one separately. Note that the term sensor grid has been used in the literature to describe a

Q. Gu, W. Zang, and M. Yu (Eds.): SEWCN 2009, LNICST 42, pp. 53–63, 2010.

sensor network with a grid like deployment topology, but such a definition does not apply to this work.

There are several rationales for sensor grids. First, the vast amount of data collected by the sensors can be processed, analyzed, and stored using the computational and data storage resources of the grid. Second, the sensors can be efficiently shared by different users and applications under flexible usage scenarios. Each user can access a subset of the sensors during a particular time period to run a specific application, and to collect the desired type of sensor data. Finally, a sensor grid provides seamless access to a wide variety of resources in a pervasive manner. Advanced techniques in artificial intelligence, data fusion, data mining, and distributed database processing can be applied to make sense of the sensor data and generate new knowledge of the environment.

Sensor grid is a relatively new area of research. Thus, the design of sensor grids is not well understood yet, unlike that of compute and data grids. Wireless sensor networks are usually based on proprietary designs and protocols, and so it is challenging to integrate them with the standard grid architecture and protocols. In this paper, we discuss the issues and challenges in the integration of wireless sensor networks with the grid.

We propose a sensor grid architecture which is based on proxy systems in order to address these design issues. The key idea is to use proxy systems as interfaces between the wireless sensor networks and the grid fabric. To study the design issues of sensor grids and improve our system design, we have developed a sensor grid testbed. This testbed consists of a set of sensor nodes or "motes", a 54-node Sun cluster based on AMD Opteron processors, and several Linux-based proxy and user systems.

2 Related Work

The Globus Toolkit [4] is becoming the de facto standard for grid middleware. It provides tools and libraries for communications, resource management, data management, security, and information services. The Global Grid Forum developed the specifications for the Open Grid Services Architecture (OGSA) [5] based on the concept of grid services. The grid services architecture enables resources to be dynamically discovered and shared. Our sensor grid architecture leverages these existing and evolving grid middleware standards and tools. Software tools for the management of wireless sensor networks are necessary for the efficient and effective utilization of wireless sensor networks. MoteLab [8] is a web-based sensor network testbed developed at Harvard University.

MoteLab, provides a web-based interface that makes it easier for users to program the motes, create sensor jobs, reserve time slots to run sensor jobs on the motes, collect the sensor data, and perform simple administrative functions. Other sensor network management software includes EmStar [9] and Kansei [10]. However, such systems can only manage a standalone wireless sensor network testbed, and they are not integrated with the grid fabric.

Recently, research efforts are beginning to study the integration of wireless sensor networks and grid computing. Researchers in the UK are studying how sensors can be integrated into e-Science grid computing applications. The Discovery Net project [22] is building a grid-based framework for developing and deploying knowledge discovery services to analyze data collected from distributed high throughput sensors.

There are also efforts to define the middleware architecture for sensor devices to facilitate the integration with the grid. The Common Instrument Middleware Architecture (CIMA) project [11] aims to "grid enable" instruments and sensors as realtime data sources to facilitate their integration with the grid. The CIMA middleware is based on current grid standards such as OGSA.

Our framework integrates wireless sensor networks with the grid by using proxy systems as interfaces between the wireless sensor networks and the grid fabric. Our architecture can support a wide range of sensor devices, even the less computationally powerful ones. Furthermore, the presented architecture is scalable, and it can integrate multiple heterogeneous wireless sensor networks with the grid.

3 Design Issues and Challenges

3.1 Grid APIs for Sensors

An approach in order integrate sensor nodes into the grid is to adopt the grid standards and APIs. The OGSA is based on established web services standards and technologies like XML, SOAP, and WSDL. If sensor data were available in the OGSA framework, it would be easier to exchange and process the data on the grid. However, since sensor nodes have limited computational and processing capability, it may not be feasible for sensor data to be encoded in XML format within SOAP envelopes, and then transported using Internet protocols to applications [12]. Grid services are also too complex to be implemented directly on most simple sensor nodes.

3.2 Network Connectivity and Protocols

In conventional grids, the network connections are usually fast and reasonably reliable. Many grid deployments leverage the Internet infrastructure. On the other hand, the sensor nodes in sensor grids are connected via wireless ad hoc networks which are low-bandwidth, high-latency, and unreliable. The network connectivity of sensor nodes is dynamic in nature, and it might be intermittent and susceptible to faults due to noise and signal degradation caused by environmental factors. The sensor grid has to gracefully handle unexpected network disconnections or prolonged periods of disconnection. Grid networking protocols are based on standard Internet protocols like TCP/IP, HTTP, FTP, etc. On the other hand, wireless sensor networks are often based on proprietary protocols, especially for the MAC protocol and routing protocol [1]. It is not practical for sensor nodes to have multiple network interface capabilities. Thus, efficient techniques to interface sensor network protocols with grid networking protocols are necessary.

3.3 Scalability

Scalability is the ability to add sensor resources to a sensor grid to increase the capacity of sensor data collection, without substantial changes to its software architecture. The sensor grid architecture should allow multiple wireless sensor networks, possibly owned by different virtual organizations, to be easily integrated with compute and data grid resources. This would enable an application to access sensor resources across increasing number of heterogeneous wireless sensor networks.

3.4 Power Management

Power management is a major concern as sensor nodes do not have fixed power sources and relies on limited battery power. Sensor applications executing on these devices have to make tradeoffs between sensor operation and conserving battery life. The sensor nodes should provide adaptive power management facilities that can be accessed by the applications. From the sensor grid perspective, the availability of sensor nodes is not only dependent on their load, but also on their power consumption. Thus, the sensor grid's resource management component has to account for power consumption.

3.5 Scheduling

In wireless sensor networks, scheduling of sensor nodes is often performed to facilitate power management and sensor resource management. Researchers have developed algorithms to schedule the radio communication of active sensor nodes, and to turn off the radio links of idle nodes to conserve power. Similarly, for applications like target tracking, sensor management algorithms selectively turn off sensor nodes that are located far away from the target, while maximizing the coverage area of the sensors. Sensor grids are data-centric in nature. A scheduler is needed for the efficient scheduling of applications to use the sensor resources for collecting sensor data. A sensor grid has a set of sensor nodes spread across multiple wireless sensor networks. These sensor nodes may provide sensors that collect different types of data such as temperature, light, sound, humidity, vibration, etc. Also, the sensor nodes may be shared by multiple applications with differing requirements.

The design of a sensor grid scheduler is influenced by some important differences between sensor jobs and computational jobs. Unlike computational jobs, sensor jobs are not multitasking in nature. A sensor node can execute only one sensor job at a time, and it cannot execute multiple sensor jobs via multitasking. While computational jobs automatically terminate upon completion, the durations of sensor jobs have to be explicitly specified. Sensor jobs are also more likely to require specific time slots for execution compared to general computational jobs.

3.6 Security

Organizations are reluctant to share their resources on a grid unless there is guarantee for security. Grid security is an active research area, in particular the security of grid services [13]. Several grid security standards and technologies have been proposed, such as the Grid Security Infrastructure (GSI) of the Globus Toolkit (GT), WS-Security [14], the Shibboleth system [23], the Security Assertion Markup Language (SAML), and the Extensible Access Control Markup Language (XACML).

Wireless sensor networks are prone to security problems such as the compromising and tampering of sensor nodes, eavesdropping of sensor data and communication, and denial of service attacks. Techniques to address these problems include sensor node authentication, efficient encryption of sensor data and communication, secure MAC and routing protocol, etc. But, for sensor grids, it is necessary to ensure that the grid security techniques and the wireless sensor network security techniques are integrated seamlessly and efficiently.

3.7 Availability

Due to the power issue and the unpredictable wireless network characteristics, it is possible that applications running on the sensor nodes might fail. Thus, techniques to improve the availability of sensor nodes are necessary.

Sensor grids should support job and service migration, so that a job can be migrated from a sensor node that is running out of power or has failing hardware to another node. If sufficient resources are available, services can be replicated so that the loss of a node will not result in service disruption. Finally, if unexpected interruptions occur, the system should be able to recover and restart the interrupted jobs.

3.8 Quality-of-Service

Quality-of-Service (QoS) is a key issue that determines whether a sensor grid can provide sensor resources on demand efficiently. Enforcing QoS in sensor grids is made complicated by the unpredictable wireless network characteristics and sensor power consumption.

The specification of the QoS requirements of sensor applications should be described in a high-level manner. A good mechanism is needed to map the high-level requirements into low-level QoS parameters. These parameters specify the amount of resources to be allocated, such as amount of sensors, memory, and network bandwidth. Similarly, service descriptions are necessary to express what a sensor service does, how to access it, and the QoS parameters of the service.

A service request might require several sensor resources. Thus, it might be necessary to make reservations of these resources to achieve the required QoS. Resource reservation is closely tied to the scheduling of sensor resources. Due to the highly dynamic sensor grid environment, any attempt at QoS provisioning should be adaptive in nature. It is necessary to consider the changes in resource availability, network topology, and network bandwidth and latency, so that the sensor grid can provide the best possible QoS to the application. Finally, mechanisms to enforce QoS in wireless sensor networks and grids have been developed separately. For sensor grids, the QoS should be enforced in a coordinated manner by integrating the wireless sensor network and the grid QoS mechanisms.

4 Architecture of Sensor Grid

4.1 Sensor Grid Organization

Figure 1 illustrates the organization of a sensor grid in terms of its resource components. A sensor grid consists of wireless sensor networks and conventional grid resources like computers, servers, and disk arrays for the processing and storage of sensor data.

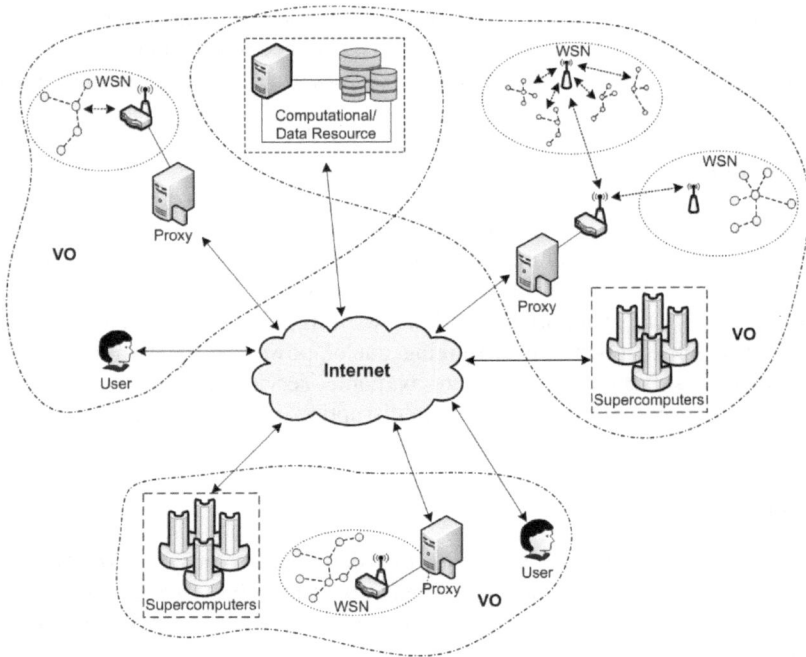

Fig. 1. Design of a sensor grid

The resources in the sensor grid are shared by several Virtual Organizations (VOs). In fact, certain resources might belong to more than one VO. Users from various VOs may access the resources in the sensor grid, even if the resources are not owned by their VOs.

4.2 Sensor Network Connectivity with Grid

We propose a proxy-based approach for our sensor grid architecture. With this approach, sensor devices can be made available on the grid like conventional grid services although they are resource constrained. Also, the proxy can support a wide variety of wireless sensor network implementations, and thus providing interoperability. Our framework is illustrated in Figure 2.

The WSN proxy in our framework acts as the interface between a wireless sensor network and the grid. The proxy serves several important functions, and addresses the design issues of sensor grids that we have discussed. First, the proxy exposes the sensor resources as grid services that can be discovered and accessed by any sensor grid application. It also translates the sensor data from its native format to a suitable OGSA format such as XML.

Second, the proxy coordinates the network connectivity between the wireless sensor network and the grid network. It provides the interface between the sensor network protocols and the Internet protocols. By using techniques like buffering, caching, and link management, the proxy can mitigate the effects of unexpected sensor network disconnection or long periods of disconnection.

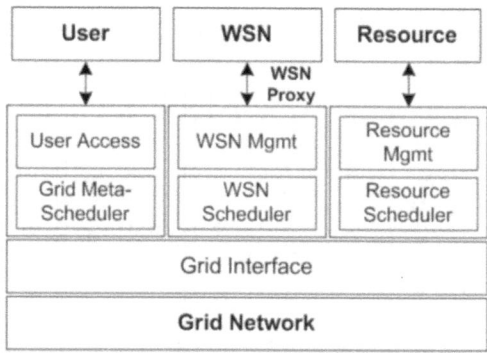

Fig. 2. Our framework

Third, the scalability of the sensor grid is enhanced by the use of the WSN Proxy. New wireless sensor networks can be integrated with an existing sensor grid by adding proxy systems. From the grid standpoint, the proxy exposes sensor resources that are accessible in a similar manner as other compute or data resources. Finally, the proxy provides various services such as power management, scheduling, security, availability, and QoS for the underlying wireless sensor network.

Our framework is based on a layered architecture approach. The layers represent the main software components that are used to build a sensor grid. Each layer defines services that are accessible via Application Programming Interfaces (APIs) for the application or other layers. The Grid Interface layer supports a standard grid middleware, such as the Globus Toolkit, that enables different types of resources to communicate over the grid network.

On the user side, the User Access layer provides an interface such as a grid portal or a workflow management tool that enables users to submit applications for execution on the sensor grid. The application might consist of sensor jobs to be executed on the wireless sensor network to collect sensor data, and also computational jobs to process the sensor data. A Grid Metascheduler, such as the Community Scheduler Framework (CSF) [15], is used to schedule and route the jobs according to their required resources.

On the wireless sensor network side, the WSN Management layer provides an abstraction of the specific APIs and protocols to access and manage the underlying heterogeneous sensor resources. It manages the configuration of the sensor nodes and provides status information of the sensor nodes. It also accepts sensor job requests from the grid and invokes the specific commands to execute the jobs on the sensor nodes.

The WSN Scheduler is the local resource scheduler for the wireless sensor network. It implements the low-level scheduling algorithms for sensor power management and resource management mentioned in Section 3.5. Furthermore, it controls the scheduling of sensor job requests from the users. The parameters of a sensor job include the amount of sensor resources required, desired start time, duration, priority, etc. The scheduler considers these job parameters, checks whether the required sensor resources are available, and reserves the resources for the job. It also works in conjunction with other Proxy Components to provide services for availability and QoS.

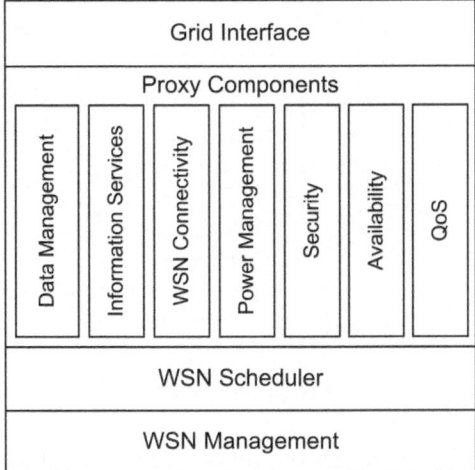

Fig. 3. Proxy software architecture

The Proxy Components work closely with the WSN Management layer to provide important services for sensor data management, information services, network connectivity, power management, security, availability, and QoS. On the resource side, the Resource Management layer provides the APIs to access and manage the computational and storage resources for the execution of grid jobs. The Resource Scheduler performs the scheduling of the grid jobs based on local usage policies.

The software architecture of the WSN Proxy is shown in Figure 3. We have already explained the Grid Interface, WSN Management, and WSN Scheduler layers in previous paragraphs. Now we will discuss the Proxy Components and the important services they provide for a sensor grid.

The Data Management component handles the conversion of sensor data from its native format to a grid-friendly format like XML or other formats desired by the user. This component also performs data fusion and other optimizations to improve the quality of data collected from multiple sensors. It supports several methods for the transfer of the sensor data to the user application, such as using GridFTP [4] or by streaming the data.

The Information Services component manages the discovery and monitoring of sensor resources. This component advertises the available sensor resources as grid services via mechanisms such as the Indexing Service (IS) of OGSA. Users can query the availability and status of sensor resources via mechanisms such as the Monitoring and Discovery Service (MDS) of OGSA. The static and dynamic information on sensor resources are directly relevant to the WSN Scheduler.

The WSN Connectivity component provides services to interface the sensor network protocols with the grid networking protocols. It buffers the transmission of sensor data, caches the routing information of sensor nodes, and manages the ad hoc sensor network links.

The Power Management component keeps track of the power consumption of the sensor nodes. Together with the WSN Scheduler, it performs actions to conserve power for the sensor nodes. The Security component implements OGSA compliant grid security technologies. For example, it can use technologies such as the Generic Security Service (GSS) API to perform authentication between the proxy and the sensor nodes.

The Availability component provides services to improve the availability of the wireless sensor network. It monitors the sensor nodes to find those with possible failing hardware or weak power level, and migrate the jobs in such nodes to the reliable nodes. To do so, it works closely with the WSN Scheduler. The Availability component can also provide fault tolerance features such as the replication of services and the recovery of interrupted jobs.

Finally, the Quality of Service (QoS) component supports the provisioning of QoS in the sensor grid. In conjunction with the WSN Scheduler, the QoS component performs the reservation and allocation of sensor resources based on QoS requirements of sensor jobs. It works with the proxy's WSN Connectivity component to adapt to the varying network conditions in order to provide the desired QoS.

5   Sensor Grid Testbed

We have developed a prototype sensor grid testbed which enables us to study the design issues of sensor grids using real hardware. This will help us to improve our framework.

5.1   Hardware Setup

The wireless sensor network in our testbed consists of 12 Crossbow MICA2 and 8 MICAz motes. The MICA2 uses a 7.3MHz Atmel ATmega128L microcontroller, 128KB of flash memory for code, 4KB of EEPROM for data, and a Chipcon CC1000 radio operating at 433MHz and 38.4Kbps data rate. The MICAz is an upgraded version of the MICA2, with the IEEE 802.15.4 compliant Chipcon CC2420 radio operating at 2.4 GHz and 250Kbps data rate.

We use 5 MIB600 Ethernet interface boards for hosting the "base station" motes. To collect sensor data, we use 12 MTS310CA sensor boards that are plugged onto the motes. Each sensor board contains a variety of sensors. The MICA2 and MICAz both run the TinyOS [16], a small open source operating system designed for embedded wireless sensor networks. The sensor applications are developed using nesC [17], an extension of the C programming language.

The testbed uses a Sun cluster with 4 server nodes and 50 compute nodes. Each server node contains dual AMD Opteron 2.2 GHz processors and 4 GB RAM, while each compute node contains dual AMD Opteron 2.2 GHz processors and 2 GB RAM. The cluster uses a Sun StorEdge 3510 FC Array with 12 TB of disk storage. We also use Linux-based PCs as the proxy systems and user systems.

5.2   Testbed Implementation

We use the Globus Toolkit (GT) 4.2 to implement the Grid Interface layer of the implemented framework. GT 4.2 is installed on the WSN proxies, the Sun cluster, and the user systems. Currently, a testbed user submits a sensor job or a computational job from the command line interface. We implemented a Grid Meta-scheduler using the Community Scheduler Framework (CSF) [15] to route sensor and computational jobs to their intended destinations.

The Sun Grid Engine (SGE) 6.2 [18] plays an important role in the implementation of the WSN Scheduler, as well as the Resource Scheduler for the compute cluster. For the WSN Scheduler, we set up a SGE queue for sensor jobs. Similarly, the compute Resource Scheduler has a SGE queue for computational jobs. To pass jobs from GT 4.2 to the SGE queues, we use a toolkit for integrating GT 4.2 with SGE called EPIC [19].

Another important component of our testbed is the WSN Management layer, which is based on MoteLab [8]. We reuse some software components of MoteLab; namely, the database backend, the job daemon, and the data logger. Our WSN Scheduler controls the operation of these components, and bypasses MoteLab's web frontend.

We also implemented additional functionalities for this layer. For example, we added more functions to monitor the status of the motes. In MoteLab, all the motes are permanently connected to the network and are programmed via MIB600 interface boards. For our testbed, five "base station" motes are connected to the network via MIB600 boards. The rest of the motes are programmed wirelessly over-the-air in an automated manner using XnP [20]. We are currently working on the design and implementation of the Proxy Components. Most of these Proxy Components have not been fully implemented yet.

6 Conclusion

Wireless sensor networks and grid computing are promising technologies that are being adopted in the industry. By integrating wireless sensor networks and grid computing, sensor grids greatly enhance the potential of these technologies for new and powerful applications. Thus, we believe that sensor grids will attract growing attention from the research community and the industry.

In this paper, we have examined the important design issues and challenges for sensor grids. To address these design issues, we proposed a novel sensor grid architecture. Also, we have developed a sensor grid testbed to study the design issues of sensor grids. From our experience, the sensor grid testbed is a very useful research tool to study sensor grid issues and to improve our sensor grid architecture design.

References

1. Akyildiz, I.F., Su, W., Sankarasubramaniam, Y., Cayirci, E.: A Survey on Sensor Networks. IEEE Commun. Mag. 40, 102–114 (2002)
2. Culler, D., Estrin, D., Srivastava, M.: Overview of Sensor Networks. Computer 37, 41–49 (2004)
3. Foster, I., Kesselman, C., Tuecke, S.: The Anatomy of the Grid: Enabling Scalable Virtual Organizations. Int. J. Supercomput. Ap. 15, 200–222 (2001)
4. Foster, I., Kesselman, C.: Globus: a Metacomputing Infrastructure Toolkit. Int. J. Supercomput. Ap. 11, 115–128 (1997)
5. Foster, I., Kesselman, C., Nick, J., Tuecke, S.: The Physiology of the Grid: an Open Grid Services Architecture for Distributed Systems Integration. Technical report, Global Grid Forum (2002)

6. Tuecke, S., Czajkowski, K., Foster, I., Frey, J., Graham, S., Kesselman, C., Maguire, T., Sandholm, T., Vanderbilt, P., Snelling, D.: Open Grid Services Infrastructure (OGSI) version 1.0. Draft Recommendation, Global Grid Forum (2003)
7. Czajkowski, K., Ferguson, D., Foster, I., Frey, J., Graham, S., Maguire, T., Snelling, D., Tuecke, S.: From Open Grid Services Infrastructure to WS-Resource Framework: refactoring & Evolution, http://www.globus.org/wsrf/specs/ogsi_to_wsrf_1.0.pdf
8. Werner-Allen, G., Swieskowski, P., Welsh, M.: Motelab: a Wireless Sensor Network Testbed. In: 4th International Symposium on Information Processing in Sensor Networks, pp. 73–78. IEEE Press, Los Angeles (2005)
9. Girod, L., Elson, J., Stathopoulos, T., Lukac, M., Estrin, D.: Emstar: a Software Environment for Developing and Deploying Wireless Sensor Networks. In: Proceedings of the USENIX Technical Conference, pp. 283–296. USENIX Association, Boston (2004)
10. Ertin, E., Arora, A., Ramnath, R., Naik, V., Bapat, S., Kulathumani, V., Sridharan, M., Zhang, H., Cao, H., Nesterenko, M.: Kansei: a Testbed for Sensing at Scale. In: 5th International Conference on information Processing in Sensor Networks, pp. 399–406. ACM, New York (2006)
11. Bramley, R., Chiu, K., Huffman, J.C., Huffman, K., McMullen, D.F.: Instruments and Sensors as Network Services: making Instruments First Class Members of the Grid. Technical report 588, Indiana University CS Department (2003)
12. Gaynor, M., Moulton, S.L., Welsh, M., LaCombe, E., Rowan, A., Wynne, J.: Integrating Wireless Sensor Networks with the Grid. IEEE Internet Comput. 8, 32–39 (2004)
13. Welch, V., Siebenlist, F., Foster, I., Bresnahan, J., Czajkowski, K., Gawor, J., Kesselman, C., Meder, S., Pearlman, L., Tuecke, S.: Security for Grid Services. In: Proc. 12th IEEE International Symposium on High Performance Distributed Computing, pp. 48–57. IEEE Computer Society, Washington (2003)
14. IBM and Microsoft Corporation. Security in a Web Services World: a Proposed Architecture and Roadmap,
    http://www.ibm.com/developerworks/library/specification/ws-secmap
15. Open Source Metascheduling for Virtual Organizations with the Community Scheduler Framework (CSF). Technical report, Platform Computing Inc. (2003)
16. Hill, J., Szewczyk, R., Woo, A., Hollar, S., Culler, D., Pister, K.: System Architecture Directions for Networked Sensors. SIGPLAN Not. 35, 93–104 (2000)
17. Gay, D., Levis, P., von Behren, R., Welsh, M., Brewer, E., Culler, D.: The nesC Language: a Holistic Approach to Networked Embedded Systems. In: ACM SIGPLAN Conference on Programming Language Design and Implementation, pp. 1–11. ACM, California (2003)
18. Bulhões, P.T., Byun, C., Castrapel, R., Hassaine, O.: N1 Grid Engine 6 Features and Capabilities. Technical report, Sun Microsystems Inc. (2004)
19. EPIC - Sun Grid Engine Integration with Globus Toolkit 3,
    http://www.lesc.ic.ac.uk/projects/epic-gt3-sge.html
20. Mote In-Network Programming User Reference. Crossbow Technology Inc.,
    http://www.tinyos.net/tinyos-1.x/doc/Xnp.pdf
21. Hui, J.W., Culler, D.: The Dynamic Behavior of a Data Dissemination Protocol for Network Programming at Scale. In: 2nd ACM Conference on Embedded Networked Sensor Systems, pp. 81–94. ACM, New York (2004)
22. Guo, Y., Darlington, J., Rueckert, D., Spence, B., Hassard, H., Cass, J., Durucan, S.: Discovery Net: an e-Science Test Bed for High Throughput Informatics,
    http://www.doc.ic.ac.uk/~mmg/dnet/testbed-Final_web.pdf
23. Shibboleth System, http://shibboleth.internet2.edu

Security for Heterogeneous and Ubiquitous Environments Consisting of Resource-Limited Devices: An Approach to Authorization Using Kerberos

Jasone Astorga, Jon Matias, Purificacion Saiz, and Eduardo Jacob

University of the Basque Country
Faculty of Engineering. Alameda de Urquijo s/n. 48013 - Bilbao
{jasone.astorga,jon.matias,puri.saiz,eduardo.jacob}@ehu.es

Abstract. Recent widespread of small electronic devices with a low capacity microprocessor and wireless communication capabilities integrated, has given place to the emergence of new communication scenarios, mainly characterized by their heterogeneity and ubiquity. As an example, in the near future, it will be very common for users to access and control electrical appliances or high performance sensors in remote locations just by using their mobile phone or PDA. However, for these environments to achieve the expected success they must probe to be secure and reliable. The security algorithms and mechanisms used to date are meant for powerful workstations and not suitable for small devices with specific constraints regarding energy and processing power. Therefore, in this paper we present a lightweight authentication and authorization solution based on the Kerberos symmetric key protocol, and we propose an extension of its functionalities in order to add authorization support.

Keywords: Authorization system, Communication system security, Kerberos.

1 Introduction

With the latest advances in semiconductor and electronic technologies and the affordability of Internet services, it is easier every day to find different electronic devices with a small microprocessor and wireless capabilities, such as Wi-Fi, Bluetooth, WiMax, ZigBee and UWB integrated. This is the case of different types of high performance sensors, electronic boards, handhelds, and even electrical appliances. Usually, these systems are characterized by small memories, limited computation power and by relying on batteries for their operation [1]. These characteristics are of vital importance when proposing a security model suitable for environments integrating this kind of devices.

An example of the considered scenarios could be the case of a user wanting to access different electronic devices or sensors at his or her home or office. For instance, in a hot summer day the user may want to connect to his or her air conditioning equipment so that when he or she arrives home the house is cool. On another occasion, the user may want to query the mobility sensor at his or her office in order to guarantee that it is currently free. Additionally, the given user could perform any of

Q. Gu, W. Zang, and M. Yu (Eds.): SEWCN 2009, LNICST 42, pp. 65–76, 2010.

these tasks from his or her mobile phone or PDA. A scenario with these characteristics can not be conceived without mechanisms that provide security. Actually, in such a case, instruments and sensors essentially become wireless physical information servers. Therefore, they must be protected in the same way as critical servers containing sensitive information or hosting vital business processes. In this sense, reliable authentication and authorization mechanisms are essential in order to avoid undesired actions to be performed by unauthenticated or unauthorized users.

A security mechanism suitable for these environments should minimize communication overhead and computation power. With these constraints it is impractical to use traditional security algorithms and mechanisms meant for powerful workstations. As a specific example, it is not practical to use asymmetric cryptosystems, and thus, key management protocols for these networks are based upon symmetric key algorithms.

Currently, the most common way to communicate with sensors or low capacity devices is by the deployment of proxies [2], [3]. The proxy is able to communicate with the sensor or low capacity device using some suitable protocol and act on behalf of this device when communicating with the external network. However, this solution presents several drawbacks: first, it is necessary to implement secure communication protocols between the end node and the proxy, and delegate the end node's credentials to the proxy so that it can act on behalf of the device. Besides this, the introduction of a new processing step in the communication penalizes the performance of the whole system. Another important drawback is that a proxy is usually designed to perform a specific task or implement a particular protocol, so such an implementation would require the development of special proxies for each application. On the other hand, recently, the efforts to introduce TCP/IP in sensors and low capacity devices are increasing considerably [4], [5].

In this paper we present a lightweight security model that provides authentication and authorization facilities in such ubiquitous and heterogeneous scenarios. For carrying out authentication this model relies on the standard Kerberos [6] symmetric key protocol. However, we propose a modification of this protocol so that it also provides authorization functionalities.

The rest of the paper is organized as follows. Section 2 provides a brief overview of the characteristics of the Kerberos protocol to show why it is a desirable mechanism for our approach and which the deficiencies that should be circumvented are. In section 3 we present an overview of different technologies that have been proposed in order to address the same issues, while in section 4 we provide a detailed description of our approach. Finally section 5 concludes the paper.

2 Kerberos-Based Approach

Kerberos is a time-tested, widely-deployed system that provides authentication and the establishment of secure channels in open networks. Each client or service in an administrative domain is called a *principal* in Kerberos, and each principal is characterized by owning a secret key known only by the principal itself and the Kerberos Key Distribution Center (KDC). The Kerberos authentication mechanism is based on the use of *tickets*. A *ticket* is a capability distributed by the Kerberos KDC that contains a proof of the identity of the principal that requested it. The tickets are encrypted so that only the entities for which they are intended are able to decrypt them. Therefore, each client that wants to authenticate to a server will present a ticket issued by the Kerberos KDC for that service. Kerberos includes mechanisms to prevent forgery of client or server identity, detect reply attacks, distribute temporary session keys for the establishment of secure channels, etc. Readers desiring a review of the Kerberos protocol are referred to [6].

Among the benefits of Kerberos that make it a suitable technology for our approach are that it prevents the transmission of passwords over the network, provides Single Sign-On functionalities and makes use of a centralized user account administration. However, Kerberos also presents some constraints that make its deployment difficult: it presents synchronization problems, due to the fact that the system clocks of all the entities that make use of Kerberos must be synchronized; and it lacks an authorization service.

The Kerberos protocol provides the mechanisms to authenticate the identity of a client to a service, but it does not provide any information about the rights of the client to access the requested service. In this situation, it is the service itself who must store and manage the privileges of the client, and implement the corresponding access control mechanisms. This fact involves also a scalability problem, because in a distributed environment consisting of different networks interconnected among them, it is not feasible that every server stores information about the privileges of each and every possible client. Especially, when this information is not static and can vary considerably during time: new users can be added, others deleted, the privileges of some users can vary, etc.

3 Related Work

Numerous efforts have been made to add authorization support to Kerberos. In fact, the designers of the Kerberos protocol already anticipated this necessity and included an optional payload field for carrying authorization information. However, the format and specification of this field was left deliberately undefined.

The Sesame [7] protocol is aimed to solve the scalability problems of Kerberos caused by the utilization of symmetric key cryptography and the lack of an access control service. This protocol is based on adding to the Kerberos KDC a new service, known as PAS (Privilege Attribute Server). Thus, when a user issues a request to the KDC it is not only authenticated, but also provided with a PAC (Privilege Attribute Certificate) which can be presented to a server when necessary.

Sesame states that it can be implemented with both public key cryptography and symmetric cryptography, but this is not completely true. Even though the authentication of the user can be carried out making use of the standard Kerberos protocol, Sesame uses public key technology in different steps. For instance, the PACs are attribute certificates which are signed by the PAS, so the PAS must own a private/public key pair. Moreover, for the inter-domain operation the KDCs make use of public key technology. In this sense, it must be taken into account that asymmetric encryption techniques are considerably more demanding than the symmetric encryption techniques from the execution time and energy consumption point of view [8], [9], and in a ubiquitous environment the energy saving can be of vital importance.

In the IDfusion [10] protocol the authors try to combine the advantages of Kerberos and LDAP with the aim of creating a system that implements authorization functionalities. IDfusion is based on the utilization of a new type of identities, known as Service Instance intrinsic identities (SIii), which represent the willingness to convey authorization to a given user for a specific service. These identities are calculated as a combination of the identity of a user (Uii) with the identity of a service (Sii), and they include a cryptographic operation using the symmetric key of the protected service. The presence or absence of a given SIii object in the directory is translated as the willingness to provide or not the access represented by that SIii value. The current implementation of IDfusion makes use of the authorization payload field of a Kerberos Service Ticket to transport the SIii value.

Although its designers claim that IDfusion is a simple system, actually it is not a lightweight protocol, due to the fact that identities are represented as N-bit vectors and cryptographic hashes, and it makes extensive use of XML. On the other hand, as SIii identities are generated as a combination of Uii and Sii identity pairs, this system presents scalability problems. In an environment with N users that can access M services in order to perform P different operations over each one, NxMxP SIii identities will have to be created and stored. Furthermore, as SIii identities are generated from individual Uii identities, authorization management can not be based on groups of users. Additionally, IDfusion does not provide any mechanism for inter-domain operation, which penalizes even more the scalability of the system.

The researches in [11] propose an authorization and accounting system which focuses on the use of *restricted proxies*. A *proxy* is defined as a kind of token that allows a given entity to act with the rights and privileges of the principal that created it. A *restricted proxy* is defined as a proxy to whom certain conditions have been imposed. In this system the concept of an authorization server is introduced, but this server does not directly specify if a given principal has the right to access a specific service or object. Instead, when the authorization server receives a request from an authorized principal, it issues a restricted proxy that allows the authorized principal to act as the authorization server with the aim of asserting the principal's rights to access certain services or objects. In each case, the authorization server includes in the proxy the actions for which the principal is authorized in the form of restrictions.

This system can be implemented with public key cryptography as well as with Kerberos. When implemented with public key cryptography, thanks to the use of certificates, a single proxy can be used to grant access to multiple services or resources. When using Kerberos, however, a proxy is actually represented by a Kerberos Service Ticket and *authenticator*, and the restrictions are introduced in the ticket's authorization payload field. Therefore, as each Service Ticket is encrypted with the secret key of the principal for which it was issued, in this case it is necessary to generate a different proxy for each of the services that are wanted to be granted access. This derives in an increased complexity and scalability problems.

Microsoft introduced Kerberos as the default authentication protocol in Windows 2000 [12]. As in Sesame, Microsoft's solution for implementing authorization is based on the concept of Kerberos PAC (Privilege Attribute Certificate). A PAC contains the principal's group membership list which is required to create the *token* used by Windows clients to make access control decisions.

The PAC is sent along with the service request messages, and it is processed by the LSA (Local Security Authority) of the resource server, which validates it and generates the client's access token, for use in subsequent authorization decisions. In the access token the LSA includes the client's authorization data found in the Service Ticket's PAC and the client's authorization data found in the local security base. Therefore, the authorization data regarding a client is not centralized in a common place where it can be easily accessed and modified by a system administrator. Instead, it is distributed among the Active Directories and domain controllers, and every application server to which the user could eventually request access.

Finally, it must be noted that these approaches to add authorization support to Kerberos, just implement the necessary mechanisms to provide the end systems with the information required to take the authorization decision. However, in all these models are, in fact, the end servers which must perform the actual authorization of the requesting clients, either by querying local access control lists or by the enforcement of other mechanisms.

4 Proposed Solution

We believe that in a heterogeneous environment where users can be added or deleted frequently and the users' rights change regularly, it is preferable to maintain all the user-related information in a centralized location, where it can be easily updated by a system administrator. Moreover, we also believe, that a system containing all the information regarding user identities and user rights is the perfect location where authentication and authorization decisions should be performed. According to this, we have modified the operation of the standard Kerberos KDC so that apart from authenticating client identities, it also verifies their user rights, and only generates Kerberos Service Tickets for those users authorized to access the desired service.

4.1 The Time Synchronization Problem

Kerberos makes use of timestamps as a way of probing the freshness of the messages, and thus avoiding reply attacks. Therefore, the system clocks of all the entities that make use of the Kerberos services must be synchronized within the allowable clock-skew window. This necessity for synchronized clocks constitutes one of the major drawbacks of using timestamps. On the other hand, the use of timestamps allows Kerberos to be stateless; state is represented through the Kerberos tickets. Statelessness is extremely valuable from the scalability point of view, especially in an environment where the majority of the targeted applications are based on simple and stateless, request-response protocols.

Given the difficulty of maintaining synchronized clocks, we propose a *nonce*-based implementation of Kerberos. As a result, the developed system becomes stateful, but it has the advantage that the state-related information is only maintained in the KDC and not in the end nodes. This fact represents an important benefit for the operation of the whole system, since a given sensor or electrical appliance can be switched off without the hitch of loosing the state-related information. Basically, this implementation uses the *authtime* field of the Kerberos tickets and protocol messages to include a nonce value which should be validated by each entity that receives such a ticket. In order to allow the validation of the received nonce values, we have introduced the concept of a *Nonce Validation Service* (NVS). The NVS is a new service that resides in the Kerberos KDC, together with the Authentication Server (AS) and the Ticket Granting Server (TGS).

The NVS maintains an information base in which each entry corresponds to a client and service principal pair and their associated nonce value. This information is updated by the TGS each time it generates a new Service Ticket with a given nonce value; and by the AS each time it generates a new Ticket Granting Ticket (TGT), which is in fact a service ticket to authenticate to the security principal representing the TGS. On the other hand, each entry has a limited lifetime, which coincides with the lifetime of the TGT or Service Ticket in which the nonce value associated to that entry was embedded. When the lifetime associated with an entry expires, the entry is deleted from the information base.

Additionally, this system allows the establishment of concurrent communications between the same principals: if multiple communications are established between the same client and service principals, there will be several entries regarding the given client/service pair in the information base maintained by the NVS.

Although we have obviated the timestamps included in the *authenticators* of the Kerberos protocol, it must be noted that these timestamps are also replaced by nonce values. In fact, they are replaced by the same nonce values embedded in the tickets the authenticators are sent with.

4.2 Authorization Information

We consider two types of security principals:
- Client principals, which represent the identities of the users that want to get access to services or resources.
- Service principals, which represent the identities of the end applications or resources that are wanted to be protected.

Regarding the service principals, it must be taken into account that many services or resources can be subject of different operations. This is the typical case of a given variable or parameter in a sensor, where the parameter can be subject of reading or writing operations. In such cases, we define the service principal identity as the operation to be performed over the final resource. Therefore, a single service will own as many service principals as operations can be performed over that service. That is, reading a variable would be considered a different service principal than writing it.

We propose an authorization solution which makes use of a role-based access control (RBAC) model [13], [14] in order to manage authorization-related information. This information is stored in an information base included in the Kerberos KDC. The information base contains two types of entries. On the one hand, it contains entries associating the client principals with a list of the roles that can be undertaken by them. There will be an entry of this type for each of the client principals defined in the Kerberos identity repository, and each client principal should be assigned at least one role. On the other hand, it contains entries associating the service principals with the different roles that should be granted access to the given service. As in the previous case, it will exist an entry of this type for each service principal defined in Kerberos;

and to each of them at least one role should be granted access, since it has no sense providing a service which no one will be able to access.

4.3 Obtaining the Service Ticket: The Authorization Decision

The authorization decision is performed by the KDC whenever a client principal requests a Service Ticket. At this point, the KDC owns all the necessary information to authenticate the requesting principal, and authorize or deny its access to the service for which it is requesting a Service Ticket. The Service Ticket request is authenticated, as in Kerberos, by the presentation of a previously obtained Ticket Granting Ticket (TGT) and the use of a session key. Once the requesting client principal has been determined to be a legitimate user, the KDC proceeds to verify if this principal owns the necessary rights to access the desired end service. With this purpose, the KDC queries its local authorization base, using as entry points the identities of the requesting client principal and the requested service principal. Then compares the roles with access right to the service with the roles that can be undertaken by the user, and if a match is found, the user is determined to be authorized to access the desired service.

In the case that a client principal is determined to be authorized to access the service principal for which it requested a Service Ticket, and only in this case, the requested Service Ticket is generated by the KDC. This way, the network load, as well as the data processing performed by the service principals, is reduced. Actually, the generation of a Service Ticket involves some resource consumption: first, bandwidth consumption, as the ticket must be transmitted from the KDC to the requesting entity and then from this entity, to the desired end server; second, CPU consumption, mainly in the resource provider, as it has to decrypt the received Service Ticket and perform the authentication and authorization of the incoming request. All these processes become a waste of resources when it can be determined beforehand that in the end the service will not be provided because the user is not authorized to access it.

Therefore, in our model, a Service Ticket conveys both authentication, as well as authorization rights for the principal that owns it. In the generated Service Tickets the authorization payload field is used to embed the identifier of the role undertaken by the client principal when accessing a resource with the present ticket. If there is more than one match between the list of roles that can be assumed by the client principal and the list of roles with access to the requested service principal, the authorization payload field will contain the identifiers of all the roles that match the condition. This information is encrypted with the secret key of the service principal for which the ticket is intended. This way it is guaranteed that only this principal can read the authorization information embedded in the ticket, and it is also avoided the possibility of third parties generating their own authorization data and sending it to the KDC as encrypted authorization data to be included in the tickets.

4.4 The Service Access Phase

When a resource provider receives a new service request accompanied by a Service Ticket, it has to validate the content and the format of the received ticket. The validity of this ticket asserts that the identity claimed by the client is true and also that the client is authorized to access the given service. Therefore, the resource providers do not need to perform further authorization checks, or implement access control mechanisms as they rely on the Kerberos KDC for both, authentication of remote users and control of unauthorized accesses.

The validation of a Service Ticket consists basically of three steps. First the ticket must be successfully decrypted with the service principal's secret key, guaranteeing this way that it was generated by a trusted KDC. Second, the validity of the nonce value contained in the ticket must be checked against the Kerberos KDC. With this purpose a request must be sent to the NVS, as shown in Fig. 1.

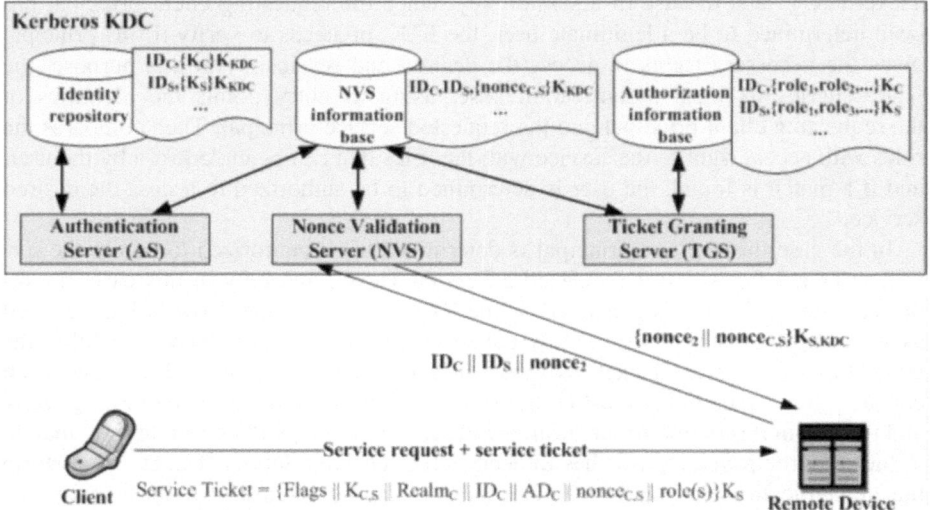

Fig. 1. Basic architecture and message exchanges of the proposed solution

In order to support the operation of the NVS, two new messages have been introduced into the standard Kerberos protocol:

- KRB_NVS_REQ: a short cleartext message specifying the identity of the client principal who sent the Service Ticket (ID_C) and the identity of the service principal that received it (ID_S). Additionally, it includes a nonce value in order to avoid reply attacks ($nonce_2$). This nonce value serves to associate the response from the NVS with the current request

$$ID_C \parallel ID_S \parallel nonce_2$$

- KRB_NVS_REP: this message just includes two nonce values, and it is encrypted with the secret key shared between the service principal and the KDC ($K_{S,KDC}$). This is not the service principal's secret key, but the session key provided by the KDC along with the TGT during the authentication process, as defined in the standard Kerberos protocol. However, in the standard Kerberos protocol, as the NVS does not exist, this key is shared only between the principal and the TGS. In the following line, the curly braces denote encryption of the information contained between them with the secret key specified just after.

$$\{nonce_2 \parallel nonce_{C,S}\}K_{S,KDC}$$

The first nonce value is the same as the one received in the request ($nonce_2$) and serves to match the response with the corresponding request. The second nonce value ($nonce_{C,S}$) is the nonce value that should be embedded in the received Service Ticket. By comparing this value with the one extracted from the Service Ticket, the service principal can determine the validity of the received ticket.

As the information included in the KRB_NVS_REQ is not authenticated nor encrypted, the NVS does not know whether this information is actually true. However, it generates a reply and sends it to the requesting entity without knowing or caring if it is really who it claims to be, or an impostor. This behaviour is acceptable because the response is encrypted with a secret session key only known by the KDC and the service principal indicated in the request. As a result, nobody but this principal will be able to decrypt and use the reply message.

Nevertheless, an attacker could send numerous KRB_NVS_REQ messages in order to get known plaintext encrypted with the session key used by some principal to communicate with the KDC, and thus attack this key. Taking into account the short length of the reply message, the number of requests sent by a malicious entity in order to perform such an attack would have to be extremely high, and thus easily detectable by any Intrusion Prevention System (IPS). Anyway, even if a malicious entity reached to get the attacked key, what it would obtain is just a temporary session key with a limited lifetime, which will be overwritten with the next TGT renewal process. As a result, the consequences of the attack would be limited.

Someone may think that conveying authorization rights by the mere issuance of Kerberos Service Tickets is not suitable as there is no way of guaranteeing that the authorization process has been actually performed. For this reason, in our approach when the TGS issues a Service Ticket, it embeds in the authorization payload field the identification of the role the user is assuming when using that ticket. The existence of this information is the third of the main validity checks that are performed by the end applications over the received Service Tickets. Additionally, this information could also be used by the end entities to perform further tasks, such as accounting.

4.5 Revocation of User Rights

Thanks to the usage of nonce values as a method to guarantee the freshness of the Kerberos tickets, the revocation of users' rights can be performed easily, just by deleting from the NVS information base the entries associated with the user and the service for which access is wanted to be denied.

When a system administrator modifies the list of roles associated with a given client principal or service principal identity, automatically all entries associated with the modified principal are deleted. This way, if a client tries to access a resource server by presenting an old Service Ticket, the nonce value included in the ticket and the one provided by the NVS during the ticket validation process will not match up, and thus, the service will not be provided. In this case, the client will have to request a new Service Ticket to the Kerberos TGT, which will perform the validation of the client's access rights, using this time the newly updated values.

However, when the modification of the authorization information occurs, the already established connections are not cut off, and they continue until the Service Ticket expires and a new one has to be requested, in which case the authorization process is performed again.

4.6 Inter-domain Operation

The development of networking technologies has made the collaboration between geographically separated enterprises and organizations easier. Thanks to these technologies organizations can easily exchange information and work together. However, for a successful data sharing and cooperative work, it is essential to implement secure and reliable communication mechanisms. User authentication and access control are two important issues in this regard.

The Kerberos protocol allows the existence of different realms or domains, each one with its own KDC. The inter-domain operation of Kerberos relies on the concepts of *referral tickets* and *inter-domain secrets*. To enable inter-domain authentication, every domain that is trusted by another domain is registered in the domain's KDC as a security principal. These principals' secret keys are usually known as inter-domain secrets. When a client principal in a given domain wants to obtain a Service Ticket for a service principal in a different domain, it has to query the remote domain's KDC. Before the client principal can contact the remote domain's KDC, it has to get from its own KDC a valid Service Ticket to talk to the remote KDC. This Service Ticket, which is encrypted with the inter-domain secret, is known as a referral ticket. It must be noted that the trust relationship between two domains may not always be direct; it can also be based on a tree hierarchy. If the trust relationship is established as a hierarchy, each domain is registered in its parent domain's KDC, and trust paths have to be followed in order to obtain a referral ticket for a domain in a different branch of the tree.

Regarding our approach, referral tickets convey, embedded in their authorization payload field, information indentifying the different roles the user is allowed to undertake in its local domain. When the remote domain's KDC receives a Service Ticket request authenticated with a referral ticket, it extracts from this ticket the list of roles the requesting principal can execute and compares them with the roles allowed to access the desired service. If a matching entry is found, the remote KDC generates a Service Ticket, including in the authorization payload field an identifier of the role the user is assuming; if not, no Service Ticket is generated, and an authorization error is returned to the requesting client principal. In order to facilitate the inter-domain operation between organizations with a collaborative relationship, some defined roles can extend over different domains. That is, the same role may be used to grant access to different services in different domains.

Thanks to these mechanisms, the Service Ticket issued by the remote domain's KDC for the remote client principal is identical to the one it would have issued for a local client principal. Therefore, when a service principal receives a request from a client of a different domain, it does not even realize that the request came from outside its domain, and it processes it exactly like any other incoming request.

5 Conclusions

Authentication and authorization are very closely related concepts. In fact, the final aim of these two processes is to guarantee that only the legitimate end users allowed to access a certain service or resource are the ones who actually access it. For this reason, we think that it is not worth providing a user with authentication credentials for a service it is not authorized to access. Therefore, we have developed a modification of the Kerberos protocol, so that before providing a user with the necessary credentials to authenticate to a service, it verifies if the user is allowed to access this service, and only in this case, generates the requested credentials. This way, users are only able to authenticate to services they are authorized to access. This supposes a resource saving for the end services, as they do not need to process authentication requests of users which in the end will not be able to access the service, as they are not authorized to do so.

We think that the best place to enforce the access control restrictions is the Kerberos authentication server. Kerberos provides the users with the necessary credentials to access a service in two steps: first, it authenticates the identity of the users and provides them with the necessary elements to establish secure and authenticated communications with the KDC. Then, it generates specific credentials for each service the users want to communicate with. It is before the second step takes place where the authorization rules should be enforced, limiting this way the generation of authentication credentials to legitimate authorized users. As a consequence, the generated credentials do not only guarantee the identity of the user they represent, but also that this user is authorized to access the service for which the credentials are intended. Thanks to this, the end application servers do not need to implement and enforce their own access control restriction, as the Service Tickets generated by the KDC already convey authentication and authorization rights for the users that own them.

On the other hand, performing authentication and authorization of users in the same server facilitates the user management processes. As all the user-related information is centralized in a single server, the network administrator only needs to interact with one server in order to add or delete users, or modify the users' access rights.

Finally, it should also be mentioned that due to the fact that it provides Single Sign-On functionalities, Kerberos is a very suitable protocol to be used in organizational environments. The user only needs to enter his or her password in the first step of the authentication process, where in fact, his or her identity is verified. From then on, the user owns all the necessary information to authenticate to the Kerberos KDC and obtain the Service Tickets needed to access all the services it is allowed to, without needing to enter the password again.

References

1. Al-Muhtadi, J., Mickunas, D., Campbell, R.: Wearable security services. In: 21st International Conference on Distributed Computing Systems, Phoenix, pp. 266–271 (2001)
2. Abadi, D.J., Lindner, W., Madden, S., Schuler, J.: An integration framework for sensor networks and data stream management systems. In: 30th international Conference on Very Large Data Bases, Toronto, vol. 30, pp. 1361–1364 (2004)
3. Kansal, A., Goraczko, M., Zhao, F.: Building a sensor network of mobile phones. In: 6th international conference on Information processing in sensor networks, Cambridge, Massachusetts, pp. 547–548 (2007)
4. Dunkels, A., Alonso, J., Voigt, T.: Making TCP/IP Viable for Wireless Sensor Networks. In: Work-in-Progress Session of the first European Workshop on Wireless Sensor Networks, Berlin (2004)
5. 6lowpan IETF group,
   http://www.ietf.org/html.charters/6lowpan-charter.html
6. Neuman, C., Hartman, S., Raeburn, K.: The Kerberos network authentication service, v5 (2005), http://www.ietf.org/rfc/rfc4120.txt
7. Kaijser, P., Parker, T., Pinkas, D.: SESAME: the solution to security for open distributed systems. Computer Communications 17(7), 501–518 (1994)
8. Ruangchaijatupon, N., Krishnamurthy, P.: Encryption and power consumption in wireless LANs-N. In: 3rd IEEE Workshop on Wireless LANs, Newton, Massachusetts (2001)
9. Potlapally, N.R., Ravi, S., Raghunathan, A., Jha, N.K.: Analyzing the energy consumption of security protocols. In: 2003 International Symposium on Low Power Electronics and Design, Seoul, pp. 30–35 (2003)
10. Wettstein, G.H., Grosen, J.: IDfusion, an open-architecture for Kerberos based authorization. In: AFS and Kerberos Best Practices Workshop, Michigan (2006)
11. Neuman, C.: Proxy-based authorization and accounting for distributed systems. In: 13th International Conference on Distributed Computing Systems, Pittsburgh, pp. 283–291 (1993)
12. Walla, M.: Kerberos explained, issue of Windows 2000 Advantage magazine (2000),
    http://technet.microsoft.com/en-us/library/bb742516.aspx
13. Sandhu, R.S., Coyne, E.J., Feinstein, H.L., Youman, C.E.: Role-based access control models. IEEE Computer 29(2), 38–47 (1996)
14. Ferraiolo, D.F., Sandhu, R.S., Gavrila, S., Kuhn, D.R., Chandramouli, R.: Proposed NIST standard for role-based access control. ACM Transactions on Information and System Security 4(3), 224–274 (2001)

Signalprint-Based Intrusion Detection in Wireless Networks

Rob Mitchell[1], Ing-Ray Chen[1], and Mohamed Eltoweissy[2]

[1] Dept. of Computer Science, Virginia Tech
{rrmitche,irchen}@vt.edu
[2] Dept. of Electrical and Computer Engineering, Virginia Tech
toweissy@vt.edu

Abstract. Wireless networks are a critical part of global communication for which intrusion detection techniques should be applied to secure network access, or the cost associated with successful attacks will overshadow the benefits that wireless networks offer. In this paper we investigate a new scheme called Nodeprints to extend the existing centralized Signalprints design for authentication to a distributed voting-based design for intrusion detection. We analyze the effect of voting-based intrusion detection designs, the probability of an individual node voting incorrectly, the ratio of mobile nodes to base stations, and the rate at which nodes are compromised, on the system performance measured by the probability that the intrusion detection system yields a false result. We develop a performance model for evaluating our Nodeprints design and identify conditions under which Nodeprints outperforms the existing Signalprints design.

Keywords: intrusion detection, wireless networks, Signalprints, identity-based attacks, performance analysis.

1 Introduction

Over the past few years various techniques have been proposed in the literature to authenticate mobile terminals in wireless networks. In particular fingerprinting techniques [1,3,4,8,9] based on side channel data [7] from the physical layer such as signal strength or signal phase and from the link layer such as timing or protocol have been developed to authenticate mobile terminals [10] Signalprints techniques [5] utilize signal metadata, specifically sequences of received signal strength indication (RSSI), collected to authenticate a reported identity and to deal with identity based attacks in wireless networks. Our work extends the existing centralized Signalprints design for authentication into a distributed voting-based design for detecting malicious nodes in wireless networks, recognizing that it is critical to apply intrusion detection techniques to secure network access, or the cost associated with successful attacks will overshadow the benefits that wireless networks offer.

Specifically, in this paper we propose, investigate and analyze a new intrusion detection design based on Signalprints, which we call Nodeprints, for securing

Q. Gu, W. Zang, and M. Yu (Eds.): SEWCN 2009, LNICST 42, pp. 77–88, 2010.

user access to wireless networks. Our Nodeprints design is fully distributed, thus eliminating the single point of attack or failure present in the Signalprints design. Moreover, we show that our Nodeprints design utilizing voting outperforms the existing Signalprints design in terms of the probability that the intrusion detection system yields a false result.

The rest of the paper is organized as follows. Section 2 surveys related work. Section 3 discusses system model, our Nodeprints intrusion detection design, performance model, as well as performance metrics used to evaluate the performance of the proposed Nodeprints intrusion detection design compared with the existing Signalprints design. Section 4 evaluates Nodeprints intrusion detection based on analysis. Finally, Section 5 summarizes the results and outlines some future research areas.

2   Related Work

Fingerprinting [1,3,4,8,9] utilizes side channel data to identify a terminal in a wireless network. The most readily-available side channel data is signal strength; this is typically measured in decibels referenced to one milliwatt (dBm). The most basic technique for fingerprinting involves a single node correlating historical signal strength data with a sample during authentication. This is a local, or host-based, signal strength approach. Patwari and Kasera enhance this basic technique by leveraging a time series of signal strength data to enhance authentication in [9]. They incorporate mobility into their design.

Desmond, et al. enhance authentication with another fingerprinting technique by analyzing 802.11 probe request frames [4]. This design associates nodes with a tuple of (Architecture, Wireless Network Interface Card Driver, Operating System). A major drawback of this approach is that it does not work with a homogeneous or near-homogeneous system. Nodes cannot be uniquely identified; they can only be categorized as having a certain configuration. A compromised node would defeat this countermeasure. This design is a passive countermeasure; it does not introduce any additional traffic into the network.

Crotti, et al. proposed a fingerprinting technique by using the premise of a traffic analysis attack as a countermeasure for authentication and availability attacks [3]. Specifically, they prosecute the size, interarrival time and sequence of datagrams to fingerprint/profile a flow of data. This is a passive technique.

Signalprints techniques deriving from fingerprinting have been proposed [5] to deal with identity based attacks specific to wireless networks. The Signalprints design uses signal metadata collected at multiple base stations. A centralized authentication server then takes those measurements to authenticate a reported identity.

Our work extends the Signalprints design to a distributed voting-based intrusion detection design for detecting malicious (or compromised) nodes to secure network access by exploring the tradeoffs between risks and rewards associated with distributed Intrusion Detection System (IDS) designs [6]. Our voting-based IDS design derives from a cooperative IDS design [2] which requires each node to preinstall a host-based IDS. Our design does not require a preinstalled host-based IDS to provide judgments if a neighbor node is compromised. Instead, as in the Signalprints design, a node utilizes physical-layer signal strength metadata as the means for detecting whether or not a neighbor node is compromised.

3 Nodeprints Intrusion Detection

3.1 System Model

While the Signalprints design [5] proposed for authentication assumes that base stations are trusted/known-good components, it treats terminals with unfettered skepticism. Like the Signalprints design, our Nodeprints intrusion detection design assumes that base stations are trusted/known-good components and treats terminals with skepticism. However, Nodeprints intrusion detection uses data from mobile nodes in the intrusion detection function. While these terminals are the target of the IDS, Nodeprints intrusion detection seeks to capitalize on the fact that not all terminals are compromised. Nodeprints is further distinguished by its voting based, distributed design. Instead of forwarding sample data, participants cast yes/no votes; this eliminates a centralized component as a single point of failure.

This threat model assumes that the adversary has some amount of physical access to the facility, but the access is not complete. In a low-tech attempt to spoof the RF fingerprint of a target, we can assume the adversary can access a hallway or office adjacent to the target. It is reasonable to assume the adversary cannot collocate precisely with the target.

This threat model also assumes that the adversary is using a discreet, unsophisticated antenna design, for example a single whip antenna. Contrast this with techniques that prosecute a much more focused attack on the physical layer by spoofing RF fingerprints with antenna arrays (collocated or geographically dispersed), parabolic designs and beamforming techniques.

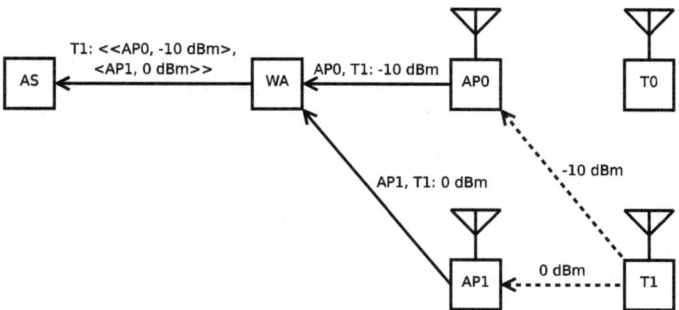

Fig. 1. Flow of information in the Signalprints design

3.2 Design

Our proposed IDS design is comprised of three functions: evaluation, detection and reaction. Figure 1 illustrates the flow of information in the Signalprints design. The "T" nodes are mobile terminals, who do not contribute data to the IDS. The "AP" nodes are access points or base stations, who are the only data contributors to the IDS in Signalprints. The "WA" node is a wireless appliance and the "AS" node is an authentication server which are discussed later.

In the Signalprints design in [5], base stations collect signal metadata and forward it to the wireless appliance (WA) which collates it (correlates metadata from independent base stations to the same authentication event). Specifically, base stations monitor the RSSI component of streams.

For detection, the WA applies "min-match" and "max-match" primitives to detect bad nodes prosecuting various authentication or availability attacks, e.g., masquerade attacks, in which a bad node impersonates a single node, or Sybil attacks, in which a bad node impersonates many nodes. The "min-match" and "max-match" primitives are implemented as $minMatches(S_1, S_2, \epsilon)$ and $maxMatches(S_1, S_2, \epsilon)$ where S_1 and S_2 are two Signalprints of interest. The output of these functions is an integer which conveys the number of positions which the Signalprints differ by at least ϵ and at most ϵ, respectively. Fewer "min-matches" and more "max-matches" correlate with greater confidence that the same terminal generated the two Signalprints which, in turn, strengthens authentication by fusing user credentials with user profile.

Figure 2 illustrates the flow of information in our Nodeprints intrusion detection design. The system selects m voters (base stations or mobile terminals) nearest to the mobile host (MH) and each will cast a yes/no vote; $m = 3$ and the MH is legitimate in this scenario. The network elects a coordinator to manage the voting process following an election protocol. The coordinator is selected randomly among the m vote-participants so that no particular node will always be the coordinator–this eliminates a single point of failure or attack. The coordinator must let all voters know each others' identities so that each voter will multicast its yes/no vote to other voters. At the end of the voting process, all voters will know the same result–either the MH is authenticated or evicted based on the majority vote. We contrast this distributed design with the single point of failure contained in the WA/AS in the Signalprints design. Note that T0 casts an erroneous vote, either accidentally or nefariously. Our analysis considers both of these situations.

While the Signalprints design uses "min-match" and "max-match" primitives to detect bad nodes in a design centralized at the wireless appliance or authentication server, our Nodeprints intrusion detection design uses a more generic voting-based algorithm which could be decentralized since the detection function of the IDS is distributed. Each vote-participant casts a yes or no vote based on the signal metadata collected with respect to the MH.

3.3 Performance Model

We develop a performance model to measure the performance of our Nodeprints intrusion detection design against the existing Signalprints design. Figure 3 lists

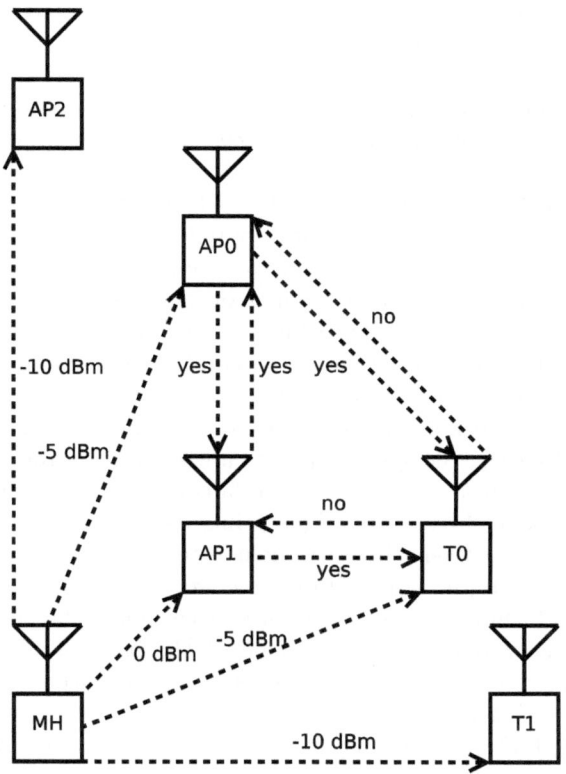

Fig. 2. Flow of information in the Nodeprints intrusion detection design

m	number of voters participating in an IDS detection event
N_{majority}	$\lceil \frac{m+1}{2} \rceil$
N_{bt}	number of bad terminals in the system
N_{gt}	number of good terminals in the system
N_{bs}	number of trusted base stations in the system
P_{fp}	probability of a false positive, a good node which the IDS detects as bad
P_{fn}	probability of a false negative, a bad node which the IDS detects as good
P_F	short for P_{fn} or P_{fp}
α	ratio of terminals to base stations
β	percentage of nodes that are captured
i	number of bad terminals voting, less N_{majority} in the special case of Figure 4
j	number of bad terminals voting in the general case of Figure 4
k	number of good terminals voting incorrectly in the general case of Figure 4
p	probability of a good node voting incorrectly

Fig. 3. Parameters

the model parameters used. For authenticating a terminal, our Nodeprints intrusion detection design involves m vote-participants selected out of those base stations and terminals reachable from a terminal node to be authenticated. If the majority out of these m vote-participants votes against the terminal node, then the terminal node is considered compromised. In our model, a *good node* is cooperative and protocol-compliant. *Bad nodes* are uncooperative (self-centered or greedy) or malicious; a bad node will always vote for a bad node and vote against a good node to facilitate attacks. When a good node casts a *correct vote*, it is the correct decision: it accurately reflects the reality of the situation. When good nodes cast *incorrect votes*, they don't do it with nefarious intentions. Rather, incorrect votes stem from error in the local IDS algorithm or the signal strength metadata that algorithm is using. A *false negative* result is one that comes from the distributed voting-based algorithm/cooperative IDS which incorrectly identifies a bad node as good. On the other hand, *false positives* mistakenly identify a good node as bad. A *false result* is a false negative or a false positive. In either case, the system fails to authenticate a reported identity. In the case of intrusion detection, the system fails to detect a compromised node.

In our model, as in the existing Signalprints design, base stations are trusted as a precondition. However, since we include terminals in intrusion detection, good nodes include both base stations and good terminals, while bad nodes include only bad terminals. The metrics used to evaluate the performance of our Nodeprints intrusion detection design against the baseline Signalprints design are P_{fn} and P_{fp}, which are the probabilities of false negative and false positive, respectively, in voting-based IDS.

The equation for P_{fn} or P_{fp} is given in Figure 4 which is the sum of two summations incorporating several system parameters, ambient conditions and indices/bounds of the model. The first parameter that system designers control is m; this is the number of voters participating in an IDS detection event. A higher value raises accuracy while a lower value economizes energy and channel overhead. N_{bs} is the other parameter that system designers control; it is the number of trusted base stations in the system. A higher value increases security

$$P_{\text{fn}} = P_{\text{fp}}$$

$$= \sum_{i=0}^{m-N_{\text{majority}}} \left[\frac{\dbinom{N_{\text{bt}}}{N_{\text{majority}} + i} \cdot \dbinom{N_{\text{gt}} + N_{\text{bs}}}{m - (N_{\text{majority}} + i)}}{\dbinom{N_{\text{gt}} + N_{\text{bt}} + N_{\text{bs}}}{m}} \right] + \sum_{j=0}^{m-N_{\text{majority}}}$$

$$\left[\frac{\dbinom{N_{\text{bt}}}{j} \cdot \displaystyle\sum_{k=N_{\text{majority}}-j}^{m-j} \left[\dbinom{N_{\text{gt}} + N_{\text{bs}}}{k} \cdot p^k \cdot \dbinom{N_{\text{gt}} + N_{\text{bs}} - k}{m - j - k} \cdot (1-p)^{(m-j-k)} \right]}{\dbinom{N_{\text{gt}} + N_{\text{bt}} + N_{\text{bs}}}{m}} \right]$$

Fig. 4. Probability of a false result under Nodeprints intrusion detection

while a lower value economizes infrastructure cost. N_{bt}, N_{gt} and p are ambient conditions: the number of bad and good terminals in the system and the probability of a good node voting incorrectly, respectively. Systems designers may know what values are reasonable to expect for these items, but they can't choose them. i, j and k act as indices and lower and upper bounds of the summations in the model. i is the number of bad terminals voting, less $N_{majority}$ in the special case of Figure 4. In the general case of Figure 4, j is the number of bad terminals voting and k is the number of good terminals voting incorrectly. The first summation is the special case; it aggregates the probability of a false result stemming from selecting a majority of bad nodes. That is, it is equal to the number of ways to choose a majority of bad nodes from the set of all bad nodes times the number of ways to choose a minority of good nodes from the set of all good nodes divided by the number of ways to choose m nodes from the set of all good and bad nodes. The second summation is the general case; it aggregates the probability of a false result stemming from selecting a majority of good nodes, some of which cast incorrect votes, coupled with selecting some number of bad nodes. That is, it is equal to the number of ways to choose a minority of bad nodes from the set of all bad nodes times the aggregate probability of a sufficient number of good nodes casting incorrect votes also divided by the number of ways to choose m nodes from the set of all good and bad nodes. The aggregate probability is a nested summation of the number of ways to choose a sufficient number of good nodes which cast incorrect votes and the remaining good nodes which cast correct votes.

4   Evaluation

In this section, we shall demonstrate improved performance of our Nodeprints intrusion detection design over the existing Signalprints design using mathematical analysis. With the baseline Signalprints design, there is only one source of error in the model: trusted base stations which vote incorrectly. With Nodeprints intrusion detection, there are three sources of error in the model: trusted base stations which vote incorrectly, good terminals which vote incorrectly and bad terminals. Our hypothesis is the increased number of terminals in the Nodeprints intrusion detection design will trade favorably with the increased error vectors in some scenarios.

4.1   Experimental Design

The effect of interest in this study is to identify design conditions under which Nodeprints intrusion detection outperforms the existing Signalprints design. There are three independent variables in this investigation: the treatment group, the capture ratio (β) and the probability of an incorrect vote (p). The first treatment group is the control group, which uses the Signalprints design; the second treatment group is the experimental group, which uses the Nodeprints intrusion detection design. The level of measurement is nominal in this case. The next independent variable, β, models the fraction of terminals that are bad. β's level of measurement is ratio; the true zero conveys no bad terminals. The final independent variable, p, models the probability that a good node votes incorrectly. p's level of measurement is ratio; no good nodes cast incorrect votes at the true zero. The dependent variable in this study, P_F is a ratio measurement. A true zero reading indicates the IDS performed correctly for the trial. N_{bs}, N_{bt}, N_{gt} and the ratio of terminals to base stations (α), which are highly covariant, are extraneous variables in this investigation. The level of measurement for the first three confounding variables, N_{bs}, N_{bt} and N_{gt} is interval. The final extraneous variable, α, models the ratio of terminals to base stations. α's level of measurement is ratio; the true zero conveys no terminals are present, only base stations.

4.2   Treatment Groups

Control Treatment. The control treatment consists of applying the existing Signalprints design to the model specified in Figure 4. Specifically, this means a configuration where $N_{gt} = N_{bt} = 0$. We exercise the independent variable p and the extraneous variable N_{bs}.

Experimental Treatment. The experimental treatment consists of applying the Nodeprints intrusion detection design to the model specified in Figure 4. We exercise the independent variables p and β and the extraneous variable α. We control N_{bs}, N_{bt} and N_{gt} by setting $N_{bs} = 3$.

4.3   Measurements

Evaluation of the false result equation shown in Figure 4 will yield measurement data. We vary the values of α, N_{bs}, p and β to test their effects. For the control treatment, pairs of p and P_F values output to a different file for each value of N_{bs}. For the experimental treatment, there are two sets of data: pairs of p and P_F values output to a different file for each value of β and pairs of p and P_F values output to a different file for each value of α.

4.4   Results

The following graphs plot the response of P_F with respect to p. In most cases, the graph of P_F is a sigmoid curve that reaches 0 and 1 as p approaches 0 and 1 respectively. This is consistent with how we expect a cumulative distribution function to behave. The exception to this is when $N_{bs} = 2$ in Figure 5.

Figure 5 presents the relationship of P_F to p in the Signalprints design for $p \in [0, 1]$ and $N_{bs} \in a_i = 2 + 4i$ where $i \in \{0, 1, 2, \ldots, 15\}$. The trend shows that the Signalprints design performs better for lower values of p and worse for higher values. The reason for this trend is that N_{bs} amplifies p. The more nodes there are in the network, the greater extent to which their incorrect vote probability will inform the result. In other words, the more data points there are in the

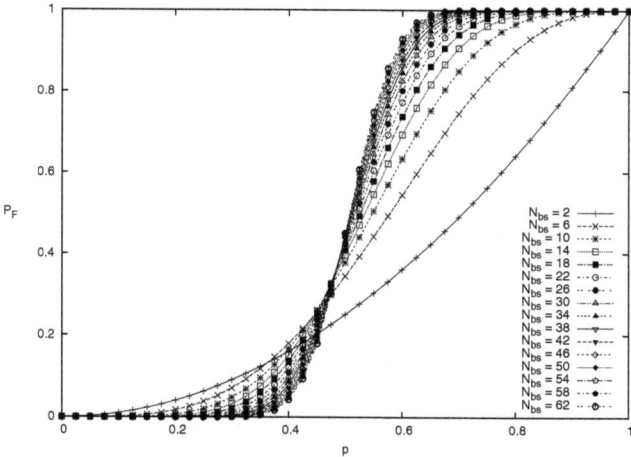

Fig. 5. P_F vs. p and N_{bs} under the Signalprints design

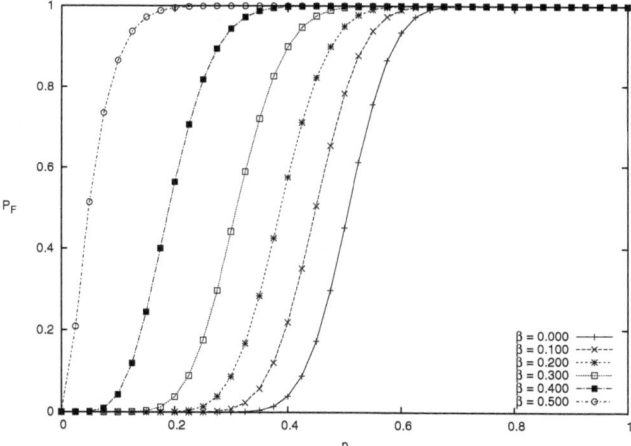

Fig. 6. P_F vs. p and β under the Nodeprints design

sample for the IDS, the more likely the system answer will adhere to the trend of individual voters.

Figure 6 conveys the relationship of P_F to p in the Nodeprints intrusion detection design for $p \in [0, 1]$ and $\beta \in \{0.000, 0.100, 0.200, 0.300, 0.400, 0.500\}$. The trend is for Nodeprints intrusion detection to perform better for lower values of β and worse for higher values. The reason behind this is that β models the node capture ratio. As more terminals are compromised, the more sense it makes not to incorporate them into the pool of voters.

Figure 7 presents the relatively small impact of α for $\beta = 0.100$. The trend is for Nodeprints intrusion detection to perform slightly better for lower values of α. The reason for this trend is that α amplifies β. The more terminals there

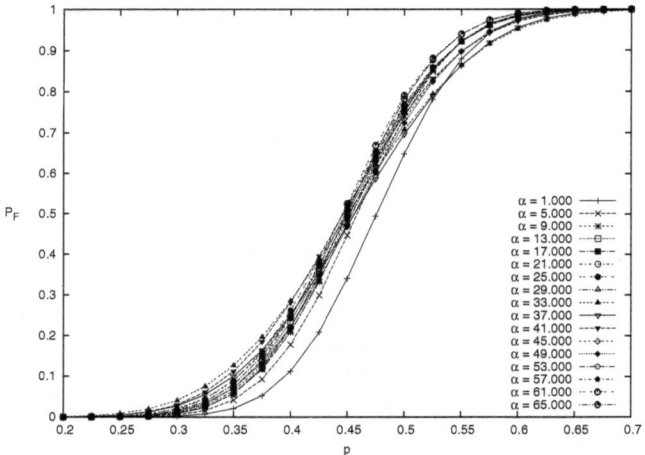

Fig. 7. P_F vs. p and α under the Nodeprints design

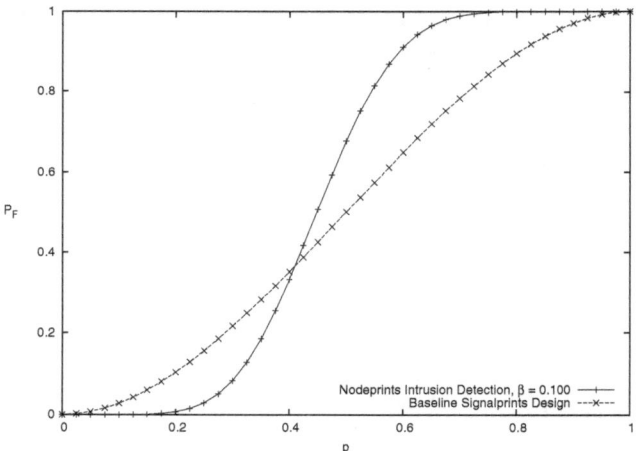

Fig. 8. Performance comparison of Nodeprints intrusion detection vs. baseline Signalprints design for $N_{bs} = 3$

are in the network, the greater extent to which their capture ratio will introduce bad voters into the pool.

Figure 8 visualizes how Nodeprints intrusion detection offers an advantage over existing Signalprints design over a wide range of parameter values under $N_{bs} = 3$ and $\beta = 0.100$. To correlate this with our model, $m = N_{bs} = 3$ for the exiting Signalprints design and $m = N_{bs} + \beta \cdot (N_{bs} \cdot \alpha) + (1 - \beta) \cdot (N_{bs} \cdot \alpha) = 15$ for Nodeprints. We observe that Nodeprints consistently outperforms Signalprints for $p \in [0, 0.400]$. It is reported [2] that any valid IDS must have its p value below 0.1. We see that for $p < 0.1$ our Nodeprints intrusion detection design significantly outperforms the existing Signalprints design.

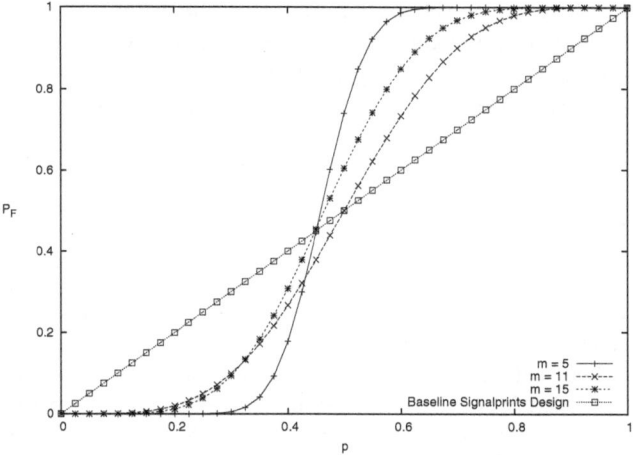

Fig. 9. Contrast Nodeprints with Signalprints with m varying for P_F vs. p

Lastly Figure 9 contrasts Nodeprints with Signalprints with m varying in the range of [5, 15]. We observe that as m increases, the cross-over value of p below which Nodeprints performs better than Signalprints increases, suggesting that Nodeprints will offer even more advantages over Signalprints when the node population is high under which the system can easily find many vote participants in a MH's neighborhood to determine whether the MH is compromised.

5 Conclusion

Signalprint-based intrusion detection contributes a novel design alternative to system designers putting together secure wireless networks. By testing the effect of system parameters such as α and β, system designers can identify situations where, when a cooperative IDS is a requirement, Nodeprints intrusion detection offers better performance than the existing Signalprints design. As a general guideline, smaller values of β and p are favorable to Nodeprints intrusion detection. Specifically, this investigation demonstrates that for $\beta \leq 0.100$ and $p \leq 0.400$, Nodeprints intrusion detection outperforms the existing Signalprints design. At the crossover point of around $p = 0.400$, the benefit of including additional metadata from untrusted terminals is overcome by the bias they impose on results.

A future investigation will consider the probability of a false result under Nodeprints intrusion detection as a function of time as terminals are identified as bad nodes by IDS and evicted from the system, and as a function of the node capture ratio. Also, future research will duplicate the results achieved in this paper in a simulation and/or experimental environment. Third, enhancing the yes/no mechanism to a yes/no/maybe or even finer granularity is worth investigating. Also, studying the trade associated with increased security and impact on network capacity or energy consumption is important. Finally, we have assumed the value of the p parameter is the same for all base stations and terminals. In heterogeneous environments this p value may vary among heterogeneous nodes, reflecting varying IDS capabilities of base stations and terminal nodes in the system. In the future we plan to test the effect of node heterogeneity on the false result probability.

References

1. Bratus, S., Cornelius, C., Kotz, D., Peebles, D.: Active Behavioral Fingerprinting of Wireless Devices. In: Proceedings of the First ACM Conference on Wireless Network Security, pp. 56–61 (2008)
2. Cho, J.H., Chen, I.R., Feng, P.G.: Effect of Intrusion Detection on Failure Time of Mission-Oriented Mobile Group Systems in Mobile Ad Hoc Networks. In: 14th IEEE Pacific Rim International Symposium on Dependable Computing, Taipei, Taiwan (December 2008)
3. Crotti, M., Dusi, M., Gringoli, F., Salgarelli, L.: Traffic Classification through Simple Statistical Fingerprinting. SIGCOMM Computer Communication Review, 5–16 (2007)
4. Desmond, L.C., Yuan, C.C., Pheng, T.C., Lee, R.S.: Identifying Unique Devices Through Wireless Fingerprinting. In: Proceedings of the First ACM Conference on Wireless Network Security, pp. 46–55 (2008)
5. Faria, D.B., Cheriton, D.R.: Detecting Identity-Based Attacks in Wireless Networks Using Signalprints. In: Proceedings of the 5th ACM Workshop on Wireless Security, pp. 43–52 (2006)
6. Frincke, D.: Balancing Cooperation and Risk in Intrusion Detection. ACM Transactions on Information and System Security, 1–29 (2000)
7. Le, T., Canovas, C., Clédière, J.: An Overview of Side Channel Analysis Attacks. In: Proceedings of the 2008 ACM Symposium on Information, Computer and Communications Security, pp. 33–43 (2008)
8. Pang, J., Greenstein, B., Gummadi, R., Seshan, S., Wetherall, D.: 802.11 User Fingerprinting. In: Proceedings of the 13th Annual ACM International Conference on Mobile Computing and Networking, pp. 99–110 (2007)
9. Patwari, N., Kasera, S.K.: Robust Location Distinction Using Temporal Link Signatures. In: Proceedings of the 13th Annual ACM International Conference on Mobile Computing and Networking, pp. 111–122 (2007)
10. Sheng, Y., Tan, K., Chen, G., Kotz, D., Campbell, A.: Detecting 802.11 MAC Layer Spoofing Using Received Signal Strength. In: INFOCOM 2008, pp. 1768–1776 (2008)

Author Index